Molecular Biology Biochemistry and Biophysics

12

T. Ando · M. Yamasaki · K. Suzuki

Protamines

Isolation · Characterization · Structure and Function

With 24 Figures

Springer-Verlag Berlin · Heidelberg · New York 1973

Toshio Ando, D.Sc.

Emeritus Professor of The University of Tokyo
and Professor of Bioorganic Chemistry,
Department of Chemistry, Faculty of Science,
Chiba University, Japan

Makoto Yamasaki, D.Sc.

Associate Professor
Department of Chemistry, College of General Education,
The University of Tokyo, Japan

Koichi Suzuki, D.Sc.

Lecturer
Department of Agricultural Chemistry, Faculty of Agriculture,
The University of Tokyo, Japan

ISBN 3-540-06221-1 Springer-Verlag Berlin Heidelberg New York
ISBN 0-387-06221-1 Springer-Verlag New York Heidelberg Berlin

 Library of Congress Catalog Card Number 73-77821. Printed in Germany. Typesetting and printing: Carl Ritter & Co., Wiesbaden. Bookbinding: Konrad Triltsch, Graphischer Betrieb, Würzburg

Preface

A century has already passed since FRIEDRICH MIESCHER, working at Strasbourg and Basel, began his study of protamine, one of the basic nuclear proteins of cells. It was first established by KOSSEL that protamine represents the simplest known protein. In the conviction that research into the nature of protamine would shed light on that of other typical proteins, a group of researchers in Germany followed MIESCHER and laid the foundations of protein chemistry. A general view of protamines was thus built up by KOSSEL, working at Strasbourg, Berlin, Marburg an der Lahn, and Heidelberg, FELIX at Heidelberg, Munich, and Frankfurt am Main, and WALDSCHMIDT-LEITZ at Prague and Munich. Concepts and techniques established by these studies have been widely utilized for research on other typical proteins.

The revolutionary advances in chemical and physical techniques after World War II extended the sphere of research to Tokyo in the Far East. Prof. FELIX' visit in 1955 greatly encouraged our research group in Tokyo. His death in August 1960 constituted a sad loss to protein chemistry and stimulated our group to assume responsibility for carrying on the studies. In the following decade we in Tokyo have been able to add a new development to the results on the chemical structure of protamines accumulated by the European researchers over a period of about fifty years.

When I was invited to write a monograph on protamines, my original intention was to add new information to the excellent descriptions by the authors I have named above, so as to provide a complete history for future reference. This I have now done with the help of my colleagues, Dr. M. YAMASAKI and Dr. Ko. SUZUKI. However, since the present authors specialize in the chemical rather than in the biological field, they acknowledge that their description of the possible biological functions of these basic nuclear proteins is unsatisfactory. The authors earnestly hope that biochemists, both in Japan and elsewhere, will work to deepen and widen this field. We shall be very happy if our book is of help to them.

The hard, but hopeful, research activity on protamines carried on in Tokyo for more than 25 years after the war was done jointly with Dr. K. IWAI, Dr. S. ISHII, and other collaborators. I do not express the least regret at having devoted so much of my life to studying the basis of the link between protamines and other proteins, because I believe that knowledge of these basic nuclear proteins will contribute to research on numerous other proteins.

I dedicate this booklet to the many senior researchers in the field of protamines. I am sincerely glad that I have been able to cooperate with them in developing the history of protamines. I am greatly indebted to Dr. J. WYMAN of the Istituto Regina Elena, Rome, for reading through this manuscript and making valuable comments. At the same time, I wish to express my deep appreciation to the editors of the series and to Springer-Verlag for giving me the opportunity to write this book.

Tokyo, September 1973 T. ANDO

Contents

Chapter I

Introduction

Nucleoproteins, in which nucleic acids combine in some manner with proteins, are found in most living cells. Modes of combination between these two groups of biopolymers appear to be of various kinds. Some are readily dissociable into their components while others are not. Examples of dissociable nucleoproteins are the nucleohistones in the nuclei of somatic cells of animals and the nucleoprotamines in the sperm cell nuclei of some fish. In these cases the nucleic acid moiety is deoxyribonucleic acid.

Studies on these nucleoproteins were started in 1868 by FRIEDRICH MIESCHER in an attempt to clarify the nature of the cell nucleus. He used materials consisting mainly of nuclei. First he found in the nuclei of pus cells nuclear protein fractions rich in phosphoric acid which he called "nucleins" (MIESCHER, 1870). Later he discovered in the sperm head of Rhine salmon a compound in which the "nuclein" was combined with a nitrogenous base in a salt-like linkage. The "nuclein" was a crude substance, later called nucleic acid. The nitrogenous base, obtained as the double salt of platinic chloride, was named "protamine" (MIESCHER, 1874, 1897). However, MIESCHER failed to observe the proteic nature of "protamine", which thus remained unknown for about 20 years.

KOSSEL and his workers obtained from the red blood corpuscles of a goose a similar salt-like compound of nucleic acid with another strongly basic protein, called "histone" (KOSSEL, 1884). Further, they found complexes of nucleic acid with various kinds of "protamine", similar but not identical to that described by MIESCHER, in the mature testes of several species of fish. Several compounds of nucleic acid with proteins resembling histone, or intermediate between protamine and histone, were also obtained (cf. KOSSEL, 1928/1929). In 1896 KOSSEL established the proteic character of the protamine now known as sturine prepared from sperm nuclei of sturgeon caught in the Baltic Sea. He based this finding on the results of the hydrolytic separation of several amino acids, including arginine, lysine, and histidine [KOSSEL, 1896 (1, 2)]. Moreover, the behavior of protamine in the presence of proteolytic enzymes was quite similar to that of the typical proteins known at that time (KOSSEL, 1898). Thus he used the generic name "protamines" to describe a group of strongly basic proteins present in the sperm cell nuclei of fish in salt-like combination with nucleic acid. He further proposed that each protamine should be named after the genus or species of the fish in which it occurred [KOSSEL, 1896 (1, 2)].

It was also KOSSEL who developed an analytical method for the basic amino acids, arginine, histidine, and lysine. His method of selective precipitation of arginine and histidine as silver salts and of lysine as a phosphotungstate salt, and the quantitative estimation of arginine, histidine, and lysine as the flavianate, picrolonate, and picrate, respectively, formed the basis of analytical protein chemistry for half a century (KOSSEL and KUTSCHER, 1900; KOSSEL, 1928).

Thus began the characterization and the analysis of the chemical structure of protamine. Protamine was considered important because of its occurrence in cell nuclei and it proved to have the simplest amino-acid composition among proteins. A large number of studies were performed over a period of more than 50 years and brilliant results were obtained by German research groups under the leadership of KOSSEL (1928/1929) in Heidelberg, FELIX (1960) in Frankfurt am Main, and WALDSCHMIDT-LEITZ (cf. WALDSCHMIDT-LEITZ and GUTERMANN, 1961, 1966) in Munich. However, they experienced difficulties arising from the lack of suitable analytical methods, from the appearance of too many arginine residues in a protamine molecule, and from the heterogeneity of protamine.

The remarkable progress made during and after World War II in chemical methods and techniques brought about a revolution in the field of protein chemistry. The development of paper and column chromatography facilitated the separation and identification of amino acids, peptides, and proteins. The introduction of the dinitrophenyl method for structural studies of proteins and the first success in determining the protein structure of bovine insulin (SANGER, 1945, 1956) were landmarks in this field. The achievement of reproducible amino-acid analysis by the ROCKEFELLER group (MOORE and STEIN, 1954; SPACKMAN *et al.*, 1958) also brought inestimable advantages to workers in the field.

The Tokyo group, making suitable use of these results, overcame the difficulties which had hampered structural studies of protamines and was able to obtain homogeneous components of clupeine and determine their complete amino-acid sequences (ANDO *et al.*, 1962; ANDO and SUZUKI, 1966, 1967; ANDO *et al.*, 1967). Thus this question was finally settled about 100 years after the term "protamine" was coined by FRIEDRICH MIESCHER. Our studies have established a general means of determining the primary structure of each component of a protamine.

The hypothesis of STEDMAN and STEDMAN (1951), which is supported by the more recent *in vitro* observations of BONNER and his colleagues [HUANG and BONNER, 1962; BONNER and HUANG, 1964; BONNER *et al.*, 1968 (1)] and ALLFREY (1966), postulates that another basic nuclear protein, histone, may modify or repress the genetic expression of deoxyribonucleic acid (DNA) in somatic cell nuclei. Thus it seems likely that the next study of biological interest based on a knowledge of the structures of the two components, DNA and protamine, will focus on the function and roles of the various components of a protamine in the sperm cell nuclei before, during, and after fertilization.

Within the scope of this book, the authors can describe only some of the features of protamines and nucleoprotamines. They would like to recommend the following reviews or books for further reading on protamines, histones, and nucleoproteins.

Select Bibliography on Protamines, Histones, and Nucleoproteins

KOSSEL, A.: The protamines and histones (Translated by W. V. THORPE). London-New York-Toronto: Longmans, Green and Co. 1928. — Protamine und Histone. Leipzig-Wien: Verlag F. Deuticke 1929.

GREENSTEIN, J. P.: Nucleoproteins. Advanc. Protein Chem. **1**, 209—287 (1944).

Ando, T., Iwai, K.: Protamines and histones (in Japanese). In: Nucleic acids and nucleoproteins (Egami, F., Ed.), Vol. I, 378—468, Tokyo: Kyoritsu Shuppan Co. 1951.

Markham, R., Smith, J. D.: Nucleoproteins and viruses. In: The proteins (Neurath, H., Bailey, K., Eds.), Vol. II, Part A, 1—122, New York: Academic Press Inc. 1954.

Davison, P. F., Conway, B. E., Butler, J. A. V.: The nucleoprotein complex of the cell nucleus, and its reactions. Progr. Biophys. Biophys. Chem. **4**, 148—194 (1954).

Ando, T., Iwai, K., Yamasaki, M.: Histones and protamines (in Japanese). In: Chemistry of proteins (Akabori, S., Mizushima, S., Eds.), Vol. III, 70—89, 89—126. Tokyo: Kyoritsu Shuppan Co. 1955.

Egami, F., Shimomura, M.: Nucleoproteins (in Japanese). In: Chemistry of proteins (Akabori, S., Mizushima, S., Eds.), Vol. III, 127—138. Tokyo: Kyoritsu Shuppan Co. 1955.

Felix, K., Fischer, H., Krekels, A.: Protamines and nucleoprotamines. Progr. Biophys. Biophys. Chem. **6**, 1—23 (1956).

Felix, K.: Protamine, Nucleoprotamine und Gene (Translated into Japanese by T. Ando and K. Iwai). In: Chemistry of proteins (Akabori, S., Mizushima, S., Eds.), Vol. V, 265—277. Tokyo: Kyoritsu Shuppan Co. 1957.

Ando, T.: Nucleoproteins (in Japanese). In: Organic chemistry (Kotake, M. and others, Eds.), Vol. 21, 428—474. Tokyo: Asakura Shoten Co. 1960.

Peacocke, A. R.: The structure and physical chemistry of nucleic acids and nucleoproteins. Progr. Biophys. Biophys. Chem. **10**, 55—113 (1960).

Felix, K.: Protamines. Advanc. Protein Chem. **15**, 1—56 (1960).

Phillips, D. M. P.: The histones. Progr. Biophys. Biophys. Chem. **12**, 211—280 (1962).

Ando, T.: Studies on the chemical structure of protamines (in Japanese). Kagaku-to-Kogyo (Chemistry and Industry) **17**, 341—357, 466—479 (1964).

Bonner, J., Ts'o, P. O. P. (Eds.): The nucleohistones. San Francisco: Holden-Day Inc. 1964.

Busch, H.: Histones and other nuclear proteins. New York: Academic Press Inc. 1965.

Murray, K.: The basic proteins of cell nuclei. Ann. Rev. Biochem. **34**, 209—246 (1965).

Vendrely, R., Vendrely, C.: Biochemistry of histones and protamines. In: Protoplasmalogia, Vol. V/3 c. Berlin-Heidelberg-New York: Springer 1966.

Hnilica, L. S.: Proteins of the cell nucleus. In: Progress in nucleic acid research and molecular biology (Davidson, J. N., Cohn, W. E., Eds.) **7**, 25—106 (1967).

Bonner, J., Dahmus, M. E., Fambrough, D., Huang, R. C., Marushige, K., Tuan, D. Y. H.: The biology of isolated chromatin. Science **159**, 47—56 (1968).

Dixon, G. H., Smith, M.: Nucleic acids and protamine in salmon testes. Progress in nucleic acid research and molecular biology (Davidson, J. N., Cohn, W. E., Eds.) **8**, 9—34 (1968).

Butler, J. A. V., Johns, E. W., Phillips, D. M. P.: Recent investigations on histones and their functions. Progr. Biophys. Molec. Biol. **18**, 211—244 (1968).

Stellwagen, R. H., Cole, R. D.: Chromosomal proteins. Ann. Rev. Biochem. **38**, 951—990 (1969).

DeLange, R. J., Smith, E. L.: Histones: structure and function. Ann. Rev. Biochem. **40**, 279—314 (1971).

Phillips, D. M. P.: Histones and nucleohistones. London and New York: Plenum Press 1971.

Hnilica, L. S.: The structure and biological function of histones. Cleveland: Chemical Rubber Co. 1971.

Chapter II

Distribution of Nucleoprotamines and Protamines

Nucleohistones are widely distributed in the somatic cell nuclei of animals, while nucleoprotamines are limited to the mature sperm cell nuclei of certain families of fish. Protamines have been separated from sperm nuclei of more than 50 species of fish and are usually named after the family of fish, according to the proposal of KOSSEL [1896 (1, 2)]. The main protamines are summarized in Table II-1 together with the genus and species names of the organisms from which they are derived. It may sometimes be necessary to re-examine the protamines listed in the table to confirm whether they are true protamines from completely mature sperm nuclei or whether they are from immature ones, which contain basic proteins of various other types, as described below.

Very few instances have been reported in which protamines or protamine-like proteins have been isolated from sperm nuclei of organisms other than fish. A basic protein, named galline, was obtained from sperm nuclei of domestic fowl *(Gallus domesticus)* and partly characterized (DALY *et al.*, 1951; FISCHER and KREUZER, 1953). It is composed of a few basic (mostly arginine) and neutral amino acids and perhaps a slight amount of glutamic acid, but has neither sulfur-containing nor aromatic amino acids except for tyrosine, and is probably of low molecular weight. It is therefore regarded as a protamine. Further studies, including fractionation into its components and some structural research, are in progress by a Japanese research group (NAKANO *et al.*, 1968, 1969, 1970, 1972) (cf. Chapter VII, B. 6).

No protamine has ever been found in animal tissues other than sperm. Although the presence of protamines has also been reported in a snail *(Helix aspersa)* [BLOCH and HEW, 1960 (1, 2)] and in lycopod pollen *(Lycopodium clavatum)* (D'ALCONTRES, 1955), these results must be subjected to careful re-examination.

Basic proteins of the histone type, or of a type intermediate between histone and protamine, are usually found in various amounts in cell nuclei of immature fish testes, often together with a protamine. Similar basic proteins of the histone type, but not protamines, are reported to be present in mature sperm cell nuclei of several fish and marine animals such as codfish (STEDMAN and STEDMAN, 1951), halibut, shad, sea urchin *(Arbacia lixula* and *Stronglocentrotus lividus)*, starfish *(Astropecten aurantiacus* and *Echinaster spositus)* (KOSSEL and STAUDT, 1926), octopus *(Eledone cirrosa)*, and some shellfish (PALAU and SUBIRANA, 1967). The same is true for some land animals such as *Drosophila* [DAS *et al.*, 1964 (1, 2, 3)], grasshopper (BLOCH and BRACK, 1964; DAS *et al.*, 1965), and cricket (KAYE and MCMASTER-KAYE, 1966). No basic proteins can be readily extracted from mammalian sperm nuclei by acids (DALLAM and THOMAS, 1953). Nevertheless, there appears to exist in mammalian semen a basic keratin-like protein containing cystine in combination with DNA in a molecular ratio

Table II-1. Origins, Classification, and Discoverers of Main Protamines

Family	Origin	Name of protamine	Classification[a]	Discovered by
Clupeidae	*Clupea harengus* (Norwegian Sea herring)	Clupeine	MP	KOSSEL, 1897; KOSSEL and DAKIN, 1904
	Clupea pallasii (Pacific herring)	Clupeine	MP	YAMAKAWA and YOSHIMOTO, 1926
	Sardinia coerulea (Californian sardine)	Sardinine	TP?	DUNN, 1926
	Amblygaster immaculatus (a kind of sardine)	Amblygine	TP	YAMAKAWA and IBUKA, 1926
Salmonidae	*Salmo salar* (Rhine salmon)	Salmine	MP	MIESCHER, 1874
	Oncorhynchus keta (Hokkaido salmon)	Salmine	MP	YAMAKAWA and YOSHIMOTO, 1926
	Oncorhynchus tschawytscha (Californian chinook salmon)	Salmine	MP	TAYLOR, 1908
	Oncorhynchus nerka (Sockeye)	Oncorhyne	TP	YAMAKAWA and NOKATA, 1923
	Salmo irideus (Trutta iridis) (Rainbowtrout)	Iridine	MP [DP (Lys)]	FELIX and MAGER, 1937
	Salmo trutta (Trutta fário) ("Bachforelle")	Truttine	MP	KOSSEL and SCHENCK, 1928
	Salvelinus namaycush (American lake trout)	Salveline	MP	KOSSEL, 1913; KOSSEL and EDLBACHER, 1913
	Salmo lacustris ("Seeforelle")	Lacustrine	DP (His)	KLEZKOWSKI, 1946
	Salmo fontinalis ("Bachsaibling")	Fontinine	MP	HAMMARSTEN, 1924
Gonorhynchidae	*Coregonus albus* (American white fish)	Coregonine	MP	KOSSEL, 1913
	Coregonus macrophthalmus ("Gangfisch" of Lake Constance)	Coregonine	MP	KOSSEL and STAUDT, 1926
Esocidae	*Esox lucius* (Pike, shad)	Esocine	MP	HUNTER, 1907
Scombridae	*Scomber scombus* (Mackerel, Baltic Sea)	Scombrine	MP	KURAJEFF, 1899
	Scomber japonicus (Japanese mackerel)	Scombrine	MP?	YAMAKAWA and YOSHIMOTO, 1926
	Thynnus thynnus (Tunny)	Thynnine	MP	ULPIANI, 1902
	Pelamys sarda (Spanish mackerel)	Thynnine	MP	KOSSEL, 1913

Table II-1 (continued)

Family	Origin	Name of protamine	Classi-fication[a]	Discovered by
	Thynnus alalonga (Germon)	Alalongine	MP	Kossel and Staudt, 1927
	Scombremorus niphonius	Scombremine	TP	Yamakawa *et al.*, 1916
	Gymnosarda vagaus	Gymnosine	TP	Yamakawa and Nokata, 1923
Xiphiidae	*Xiphias gladius* ("Schwertfisch")	Xiphiine	MP	Kossel, 1913
Sciaenidae	*Sagenichthys ancylodon* (South-American)	Ancylodine	MP	Kossel and Staudt, 1927
	Sciaena schlegeli	Sciaenine	TP	Yamakawa *et al.*, 1916
Cyclopteridae	*Cyclopterus lumpus* (Gurnard, Baltic Sea)	Cyclopterine	MP	Markowin, 1899
Tetra-odontidae	*Spheroides rubripes*	Spheroidine	MP	Yamakawa *et al.*, 1923
	Spheroides pardalis	Spheroidine II	TP	Yamakawa *et al.*, 1923
Mugilidae	*Mugil cephalus* (Mullet)	Mugiline		Yamakawa and Nokata, 1926
	Mugil japonicus (Formosan grey mullet)	Mugiline β	MP	Hirohata, 1929
Holocen-tridae	*Perca flavescens* (Yellow perch, North America)	Percine	DP (His)	Kossel, 1913
	Stizostedium vitreum (Pike perch, North America)	Percine		Kossel, 1913
	Luciperca sandra	Percine		Lisitzuin and Aleksandrovskaya, 1933
Serranidae	*Lateolabrax japonicus*	Lateoline	TP	Yamakawa *et al.*, 1916
	Stereolepis ishinagi (Sea-bass)	Stereoline	TP	Yamakawa *et al.*, 1916
Labridae	*Crenilabrus pavo*	Crenilabrine	DP (Lys)	Kossel, 1910
Cyprinidae	*Cyprinus carpio* (Carp)	Cyprinine	DP (Lys)	Kossel and Dakin, 1904
	Barbo fluviatilis ("Flussbarbe")	Barbine	DP (Lys)	Kossel and Schenck, 1928
	Leuciscus rutilus (Barbel, "Rotkarpfen")	Leuciscine	TP	Kossel and Staudt, 1927
Acipen-seridae	*Acipenser sturio* (German sturgeon, Baltic Sea)	Sturine	TP	Kossel, 1896

Table II-1 (continued)

Family	Origin	Name of protamine	Classification[a]	Discovered by
	Acipenser stellatus (Sturgeon, Caspian Sea)	Sturine		KURAJEFF, 1901
	Acipenser guldenstädtii (Sturgeon, Caspian Sea)	Sturine		MALENÜCK, 1908
	Acipenser huso (Russian sturgeon)	Acipenserine		LISITZUIN and ALEKSANDROVSKAYA, 1933
Scombropidae	*Scombropus boops*	Scombropine	TP	YAMAKAWA *et al.*, 1916
Carangidae	*Seriola aureovittata* (Yellowtail)	Serioline	TP	YAMAKAWA *et al.*, 1916
Sparidae	*Pagrosomus major* (Red sea-bream)	Pagrosine		YAMAKAWA *et al.*, 1923
Lutianidae	*Lutianus vitta*	Lutianine	MP?	YAMAKAWA *et al.*, 1923
Exocoetidae	*Cypselurus agoo* (Flying fish)	Cypseline	TP	YAMAKAWA *et al.*, 1923
Drosomatidae	*Konosirus punctatus*	Konosirine		YAMAKAWA *et al.*, 1923
Pleuronectidae	*Limanda angustirostris* (Dab)	Limandine		GOTO, 1907
	Hippoglossus hippoglossus (Halibut)	Hippoglossine		YAMAKAWA and NOKATA, 1926
Phasianidae	*Gallus domesticus* (Domestic fowl)	Galline		DALY *et al.*, 1951

This table was reproduced with some changes in arrangement from those given in the following publications. ANDO, T., IWAI, K., in: Nucleic acids and nucleoproteins (EGAMI, F., Ed.). Vol. I, p. 384. Tokyo: Kyoritsu Shuppan Co. 1951; and ANDO, T., IWAI, K., YAMASAKI, M.: In: Chemistry of proteins (AKABORI, S., MIZUSHIMA, S., Eds.), Vol. III, p. 90. Tokyo: Kyoritsu Shuppan Co. 1955.

[a] MP denotes a monoprotamine, DP(His) or DP(Lys), a diprotamine containing as basic amino acids arginine and histidine or arginine and lysine, and TP a triprotamine containing all the three basic amino acids.

of basic residues/P = 0.9 to 1.0 as determined by chemical studies (BRIL-PETERSEN and WESTENBRINK, 1963; HENRICKS and MAYER, 1965). During spermatogenesis in mouse (MONESI, 1964, 1965), rat (LINSON, 1955; VAUGHN, 1966), and bull (GLEDHILL *et al.*, 1966), basic proteins of the histone type are found. No protamines are found in mammals. Recently, however, a basic protein has been isolated from bull sperm heads (COELINGH *et al.*, 1969). It resembles protamine in its amino-acid composition though it contains in addition six half-cystine residues and numbers of tyrosine, phenylalanine, histidine, and glutamic acid residues (see Chap. VII, B. 7).

There have thus been found in the nuclei of animal cells various types of basic nuclear proteins, ranging from more or less basic histones to strongly basic protamines, including some intermediate among the two groups. It is possible that they modify the gene action of DNA to a varying extent and that the differences in function might be reflected in their structural features, *i.e.*, basicity, hydrophobicity, and conformational rigidity.

From this point of view, it is no longer reasonable to set a boundary between protamines and histones: both belong to a group of basic nuclear proteins associated at some time with DNA in sperm or somatic cell nuclei of animals (MURRAY, 1964). It is not yet clear whether protein components other than protamine are present in completely mature sperm cell nuclei of fish. FELIX stated that all the sperm nuclei of fish he had studied consisted exclusively of nucleoprotamines (FELIX, 1952, 1953). POLLISTER and MIRSKY stated that nucleoprotamine accounted for 91% of dried and defatted sperm nuclei of trout (POLLISTER and MIRSKY, 1946; MIRSKY and POLLISTER, 1946). In contrast, STEDMAN and STEDMAN emphasized that only about 68% of such sperm nuclei of salmon was protamine in combination with DNA, the remainder being a crude complex of nucleic acid with a rather acidic protein of lower arginine content called chromosomine [STEDMAN and STEDMAN, 1947 (1, 2)]. Recently a small amount of an acid-insoluble protein was found besides protamine in trout testis during spermatogenesis (MARUSHIGE and DIXON, 1969). A similar non-basic protein (acid-insoluble) was isolated from mature sperm cell nuclei of Pacific herring associated with some 14% of the total DNA in the nucleus, the rest of the DNA being mostly combined with protamine (KAWASHIMA and ANDO, 1969; KAWASHIMA, 1970). If such additional protein is always present in completely mature sperm cell nuclei as in somatic cell nuclei, the function of protamines in the cell nuclei before and/or after fertilization may have to be considered in relation to such a protein complex.

Chapter III

Preparation of Nucleoprotamines and Protamines

The starting material used for preparing both nucleoprotamines and protamines is sperm heads or testis chromatin of fish obtained from freshly collected mature milts or sperm freshly drawn from mature testes. These mature milts or sperm contain essentially only one type of cell, spermatozoa, although the former often contain immature sperm cells and connective tissues which must be removed. Ways of preparing sperm heads, chromatin, nucleoprotamines, and protamines will be briefly reviewed in this chapter[1].

A. Isolation of Sperm Heads

The preparation of sperm heads from fish is fairly easy, because the spermatozoa of fish are apt to lose their tail fibrils on stirring or homogenizing in an appropriate aqueous solution, leaving the sperm heads intact. With progress in methods of isolating cell nuclei, several modifications have been added to the original procedure of Kossel (1929).

1. Outline of the "Classical" Procedure of Kossel

A suspension of ground milts in water is filtered through a cotton cloth. Acetic acid is added to the filtrate to bring the pH of the suspension acidic to Congo red and to coagulate the sperm heads, and the precipitate is collected by centrifugation. Sperm heads thus collected are washed successively with dilute acetic acid, water, acetone (or ethanol) and ether, and finally dried over phosphorus pentoxide. The yield of dried defatted sperm heads, somewhat coloured indicating contaminations mentioned below, is usually 10—15% of the ripe milts [Ando *et al.*, 1957 (3); see also modifications by Rasmussen, 1934, and by Felix and Mager, 1937). The procedure is simple and can be used for routine work. The defect of this procedure is that various kinds of proteins from body fluid in the testis are also precipitated, being adsorbed on the sperm heads, and the specimen is also contaminated by fine debris from the testes.

2. Isolation of Sperm Heads by Fractional Centrifugation in Isotonic Salt Solution

Ripe testes or sperm from mature testes are homogenized in isotonic salt solution. The suspension is filtered through cheesecloth and the resulting suspension is repeatedly centrifuged at higher (4,000 r.p.m. for 10—5 min) and lower (1,300 r.p.m. for 2 min) speed to remove the body fluid and to exclude the debris from the testes. The snow-white precipitate of sperm heads is finally defatted and dried as described above. Examples of successful application of this procedure to the frozen testes of herring (Yamasaki, 1958) using 0.05 M sodium citrate solution (Mirsky and Pollister, 1946) and to the testes or sperm from

[1] As with proteins, all procedures should be carried out at a temperature below 5 °C unless otherwise stated.

rainbow trout [ANDO and HASHIMOTO, 1958 (1)] using Ringer's solution for freshwater fish or 0.14 M sodium chloride (cf. also POLLISTER and MIRSKY, 1946; DALY *et al.*, 1951; FISCHER and KREUZER, 1953) have been reported.

3. "Plasmolysis" Procedure of Felix

FELIX *et al.* [1951 (1, 2)] have reported successful application of "plasmolysis" to fish spermatozoa: *e.g.*, trout sperm are suspended in a large excess of distilled water and homogenized for about 1 h. During the operation, the tail fibrils and swollen cytoplasm are removed from the sperm heads. The suspension is centrifuged and the procedure repeated once more with the precipitate. The white precipitate obtained is treated with 0.005% citric acid and the precipitate collected by centrifugation is defatted and dried as described above.

This procedure gives pure white nuclei and seems to be suitable for fish spermatozoa having a low content of cytoplasmic materials and fragile tail fibrils, although the yield of nuclei is reduced by loss of nuclei in suspension into the supernatant solution.

4. Isolation of Sperm Heads and/or Cell Nucleus Fraction or Chromatin from Fish Testes during Spermatogenesis

ANDO and HASHIMOTO [1958 (1, 2)] have reported successful isolation of sperm heads and/or nucleus fraction from the testis of rainbow trout during spermatogenesis by applying the procedure described in Section 2.

In recent studies on the isolation and characterization of various nuclear proteins from both maturing and mature trout testes, the Tokyo group (KOGA *et al.*, 1969) prepared the testis cell nucleus fraction by centrifugation through 2.2 M sucrose solution, essentially according to the method of GEORGIEV *et al.* (1960) for somatic cells.

Testis cells of the rainbow trout were homogenized in 2.2 M sucrose solution containing 0.05 M sodium glycerophosphate at pH 7.9 The mixture was filtered through layers of gauze and centrifuged at 14,000 × *g* for 20 min. The treatment was then repeated, the sediment was blended with 0.25 M sucrose solution and centrifuged. The purified nuclear fraction thus obtained was used as material for further preparation of basic and acidic nuclear proteins (cf. Chap. III. C. 2. c)γ).

For their study of the developmental changes in chromosomal composition and template activity during maturation of trout testis, MARUSHIGE and DIXON (1969) applied the method of MARUSHIGE and BONNER (1966) for liver chromatin and obtained testis chromatin with a consistent recovery of over 80% relative to DNA content.

Frozen testes (0.2—2 gm) were ground using a Waring blendor in 0.075 M NaCl'$_M$ 0.024 M EDTA solution of pH 8. The sediment was washed with the same saline-EDTA solution and then with 0.01 M Tris buffer at pH 8 by repeated resuspension and centrifugation. The crude chromatin was then purified by centrifugation through 1.7 M sucrose containing 0.01 M Tris buffer at pH 8. The chromatin pellet thus obtained was washed with the Tris buffer by centrifugation at 17,000 × *g* for 20 min and resuspended in the 0.01 M Tris buffer at pH 8. The chromatin thus purified had an ultraviolet absorption spectrum typical of isolated chromatins [BONNER *et al.*, 1968 (2)].

B. Isolation and Purification of Nucleoprotamine

As with nucleohistones, nucleoprotamines can be extracted and purified by utilizing two properties of these complexes: they are soluble in high salt concentrations (usually alkali halide), in which the ionic bonding between DNA and the basic protein is broken due to double decomposition; and they have minimum solubility in isotonic salt solutions (cf. HUISKAMP, 1901). For this purpose, freshly prepared wet sperm heads are extracted by homogenization in a concentrated (1—2 M) NaCl solution; the resulting viscous solution is centrifuged to remove insoluble materials and the supernatant is poured into water to give a 0.14 M NaCl solution. The resulting fibrous precipitate of nucleoprotamine can be further purified by repeating the procedure, or can be dried by treating it with ethanol and ether [MIRSKY and POLLISTER, 1946; WATANABE and SUZUKI, 1951 (1, 3); SUZUKI and WATANABE, 1952 (1)]. With care it is possible to prepare a nucleoprotamine specimen with the DNA/protamine ratio more or less identical to that in the original spermheads (see Chap. IV). However, it must be kept in mind that, in the precipitated specimen, the structure of both DNA and protamine and the mode of binding between the two are not necessarily the same as in the intact material.

It is known that nucleohistone is to some extent soluble in pure water at 5 °C (CARTER and HALL, 1940) or in a very dilute salt solution (ZUBAY and DOTY, 1959; 0.7 mM potassium phosphate buffer, pH 6.8) in an undissociated form (CARTER, 1941; ZUBAY and DOTY, 1959) and can be extracted by careful homogenization with pure water after adequate pretreatment of material. However, it is not clear with nucleoprotoamine whether the same procedure can be applied to the extraction and purification of the complex with high yield and with reproducibility.

C. Separation and Purification of Protamines from Sperm Heads or Nucleoprotamine

In this section the development of the procedures to separate and purify protamine from sperm heads or nucleoprotamine, prepared as described in the preceding sections, will be reviewed historically according to the principle of the procedures.

1. Classical Procedures

The classical procedures of KOSSEL (1929) for the preparation of protamines may be divided into two parts (a and b) according to the nature of the reagents used to extract sperm heads.

a) Extraction with Dilute Mineral Acid

Sperm heads are treated with dilute mineral acid (hydrochloric or sulfuric) (KOSSEL, 1929). By adding a solution of anion such as picrate to the acid extract, protamine is precipitated as an insoluble salt, which is further converted into sulfate or hydrochloride. In the original method of preparation of a protamine from testes of Rhine salmon, MIESCHER (1897) used 1—2% hydrochloric acid for the extraction and platinum chloride for the precipitation of protamine. KOSSEL (1929) used 1% sulfuric acid while STEDMAN and STEDMAN (1951) used 0.1—0.5 N sulfuric acid in their "complete extraction procedure". BLOCK *et al.* (1949) isolated the sulfate salt of clupeine and salmine on a larger scale, using 0.2 N hydrochloric acid and metaphosphate, respectively, as the extracting and precipitating reagents.

Rasmussen's procedure (RASMUSSEN, 1934) which involves the extraction of sperm heads four times with 10-fold (v/w) volume of 0.2 N hydrochloric acid for 45 min at 0 °C with subsequent addition of 1/8 M sodium picrate to precipitate the picrate salt of clupeine, was routinely used by the Tokyo group (ANDO *et al.*, 1957; YAMASAKI, 1958). In this procedure, the risk of decomposition of both the protamine and DNA moieties is minimized by the brevity of contact of the material with rather low concentrations of acid at low temperature.

The protamine picrate salt is then usually converted to the sulfate salt as follows. The picrate salt is dissolved in 67% (w/w) aqueous acetone (in which the picrate salt of adenine derived as a decomposition product from DNA remains insoluble), and sufficient 2 N sulfuric acid to bring the pH to 2 and ethanol are added to the solution to precipitate the crude sulfate salt. The crude protamine sulfate is further purified by dissolving it in a minimum amount of water and precipitating the sulfate salt from a weakly acidic solution by adding ethanol. The purification procedure used to be rather tedious, because the contaminant picrate could hardly be removed by repeated reprecipitation. This difficulty was overcome by the use of an anion-exchange resin, such as Dowex 2 (sulfate form); a single passage of the aqueous solution of the crude sulfate salt through the ion-exchange column effects the complete removal of the contaminating picrate (SAWADA, 1959).

Protamine sulfate can easily be converted to the hydrochloride salt by the use of anion-exchange resins: an aqueous solution of protamine sulfate is applied to a column packed with Amberlite IRA-400 (chloride form), the column is eluted with water and the effluent fractions containing protamine are lyophilized to give the hydrochloride salt (ANDO *et al.*, 1957; YAMASAKI, 1958). By treating the picrate salt with 0.2 N hydrogen chloride-anhydrous methanol at room temperature, FELIX's group obtained protamine methyl ester hydrochloride salt (RAUEN *et al.*, 1952).

b) Extraction with Aqueous Cupric Salt

To avoid the use of mineral acids, KOSSEL and SCHMIEDEBERG (see KOSSEL, 1929, p. 19) devised the use of cupric chloride (or sulfate) to extract protamine from sperm heads. Dry sperm heads are treated with aqueous cupric chloride (or sulfate) for several days at room temperature. During the process, due to double decomposition, the protamine moiety is transferred into the aqueous phase in the form of hydrochloride (or sulfate) salt, while the nucleic acid moiety is converted to the insoluble cupric salt; after removal of the insoluble salt, the protamine fraction is precipitated from the solution as picrate salt.

The procedure has been applied in recent years to the preparation of clupeine from *Clupea harengus* (WALDSCHMIDT-LEITZ and GUDERNATSCH, 1957), protamines from several species of Salmonidae (WALDSCHMIDT-LEITZ and GUTERMANN, 1961), and thynnine from *Thynnus thynnus* (WALDSCHMIDT-LEITZ and GUTERMANN, 1966). HIROHATA's group used a similar procedure of NELSON-GERHARDT (1919) to prepare and purify mugiline β from *Mugil japonicus* (OTA *et al.*, 1959; OTA, 1961). Using cupric chloride, ANDO's group also used this procedure with success to precipitate the nucleic acid fraction from a very viscous solution of nucleoprotamine in 2 M NaCl and to recover the protamine fraction from the supernatant as a precipitate in the form of a picrate salt which is converted finally to the sulfate (ANDO *et al.*, 1957; YAMASAKI, 1958).

c) Further Purification of Protamine Preparations

A crude preparation of whole protamine can be further purified by repeating the precipitation of the sulfate salt from an aqueous solution (containing a few drops of 2 N sulfuric acid) with the addition of several volumes of ethanol. Protamine sulfate can also be purified by utilizing the difference in saturation in aqueous solution of the sulfate salt at room temperature and at 0 °C: on cooling a solution which was saturated at higher temperature, protamine sulfate usually deposits as an oil which is collected to give a purer specimen (KOSSEL, 1898; WALDSCHMIDT-LEITZ and VOH, 1954).

Because of the resistance of clupeine and protamines from several kinds of Salmonidae to the action of pepsin (see KOSSEL and MATHEWS, 1898; ROGOZINSKI, 1912; PORTIS and ALTMAN, 1947; YAMASAKI, 1960), WALDSCHMIDT-LEITZ's group employed crystalline pepsin to remove nonprotamine-proteinous impurities from these protamine preparations (WALDSCHMIDT-LEITZ and GUDERNATSCH, 1957; WALDSCHMIDT-LEITZ and GUTERMANN, 1961, 1966; see also the similar procedure of NELSON-GERHARDT [1919] used for the purification of salmine). Obviously proteolytic enzymes such as pepsin should be used with great care in the purification of protamines, and their use is limited even within this group of proteins.

2. Modern Improvements to Classical Procedures

Since the 1940s and 1950s, with the recognition of the heterogeneous nature of protamines (see Chap. VII) and with the introduction of new materials and techniques in protein chemistry, such as ionexchange, countercurrent distribution and gel filtration, various improvements have been made to the classical procedures, permitting the preparation of protamines in a milder way and without fractionation.

a) POLLISTER and MIRSKY (1946) tried to separate protamine and DNA by dialyzing a solution of nucleoprotamine (from *Salmo fario*) in 1 M NaCl in a Visking tube against a saline solution of the same concentration. Protamine hydrochloride with a molecular weight of several thousands was recovered in the diffusate, while the sodium deoxyribonucleate with highmolecular weight remained in the tube as retentate.

b) FELIX and his colleagues prepared whole clupeine hydrochloride in the two following ways:

(α) Dry, fibrous nucleoiridine (1.5 g) is ground in a mortar with quartz sand. The finely ground material is extracted with several small portions of cold 0.2% HCl. The combined extracts are centrifuged to obtain a clear supernatant (150 ml). To the supernatant is added 20 volumes of cold acetone and the resulting cloudy mixture is left for 12 h at 4 °C. During that time, iridine hydrochloride deposits on the wall of the flask as a thin, transparent precipitate. The acetone layer is discarded and the precipitate is placed in a centrifuge tube and dissolved in a minimum amount of water. Acetone is then added to precipitate the protamine from the solution. The precipitate is collected by centrifugation and lyophilized to give a hard, crystal-like powder of iridine hydrochloride (yield 200 mg) [FELIX *et al.*, 1951 (1)].

(β) Nucleoclupeine (100 mg) is dissolved in cold 10% sodium chloride by stirring. To the clear but somewhat opalescent solution are added crystals of sodium chloride until saturation is reached, when the supernatant immediately turns to a milky mixture. The suspension is centrifuged at 20,000 r.p.m. for 1 h to collect the salted-out clupeine hydrochloride. The precipitate, which is a mixture of clupeine hydrochloride and excess sodium chloride crystals, is dissolved in 5% trichloroacetic acid and the solution is filtered to remove the slightly brown, fibrous precipitate of contaminated deoxyribonucleic acid. To the clear filtrate is added, dropwise, 20% trichloroacetic acid until the solution appears milky (approximately 8% of trichloroacetic acid in the final solution). After standing for several hours, the clupeine fraction settles to the bottom of the vessel as an oil. After removal of the supernatant, the oil is dissolved in water and lyophilized to give clupeine trichloroacetate salt (yield 40 mg). Alternatively, the oily deposit of the trichloroacetate salt is taken in 0.2% hydrochloric acid and excess acetone is added to precipitate the hydrochloride, which is then dissolved in water and lyophilized to give clupeine hydrochloride (GOPPOLD-KREKELS and LEHMANN, 1958).

c) Ando's group in Tokyo has used the following procedures *(α)* and *(β)* to obtain whole protamine minimizing fractionation during the process of preparation, and (γ) to study the composition of basic nuclear proteins (histones and protamine) during spermatogenesis:

(α) Sperm heads (1.02 g) of *Salmo irideus* prepared according to the procedure of RASMUSSEN (1934) (a modification of KOSSEL's method; see Section A. 1) or by the fractional centrifugation method (Section A. 2), are extracted four times with 10 ml each of cold 0.2 N hydrochloric acid; each extract is neutralized with Amberlite IRA-400 HCO_3 form). On lyophilization, white amorphous powder of whole iridine hydrochloride is obtained. The yields are 166 mg from the first extract, 73 mg from the second, and 55 mg from the combined third and fourth extracts (ANDO and SAWADA, 1960). The same authors similarly prepared clupeine hydrochloride from sperm heads of specimens of *Clupea pallasii* caught on three different coasts of Hokkaido, with consistent yields of 28 to 30 % (ANDO and SAWADA, 1962).

(β) The following procedure for the purification of protamines utilizing absorption to and elution from a column of Amberlite CG-50 [ISHII, 1960 (2)] is used routinely by the Tokyo group.

For example, dry sperm heads (20 g) from *Oncorhynchus keta* are ground in a mortar with powdered quartz (10 g) and 0.2 N HCl (5 ml) in a cold room. The mixture is extracted several times with cold 0.2 N HCl. The combined clear extracts are neutralized with dilute NaOH to pH 4.0 and applied on an Amberlite CG-50 column equilibrated with an acetate buffer of pH 4.0. The column is washed with cold 0.2 M acetic acid until the effluent fractions give a negative response to the SAKAGUCHI test. This procedure does not remove salmine but some histone-like basic proteins giving the positive color reaction, if present, can be washed out from the column (the sodium ion required to equilibrate the column is also washed out by the procedure). The protamine fraction on the column is then eluted with cold 0.05—0.1 N HCl; the effluent fractions with the positive SAKAGUCHI test are combined, neutralized to pH 4.0 with Amberlite IRA-400 (HCO_3 form), and lyophilized to give whole salmine hydrochloride. Otherwise, salmine in the acid effluent fractions is recovered as the sulfate salt (yield 4.5 g, 22.5 %) by treating the effluents with ethanol and sulfuric acid (WATANABE, 1969).

(γ) To study nuclear proteins during spermatogenesis of rainbow trout, KOGA *et al.* (1969) separated the basic nuclear protein fraction by employing essentially the method of WANG (1967) for somatic cells.

The testis cell nuclear fraction purified by 2.2 M sucrose centrifugation as described above (see Section A. 4) is homogenized in 5 volumes of 0.14 M sodium chloride solution for 40 min in a Teflon homogenizer. After centrifugation at 15,000 × *g* for 10 min the solution of globular protein fraction is removed and the residue is treated similarly once again. The residue is then suspended in 50 volumes of 0.14 M NaCl solution and an equal volume of 2.86 M saline is added to bring the concentration of saline in the mixture to 1.5 M. The mixture is then left overnight with stirring at 4 °C. After removal of the residual fraction (residual proteins and probably ribonucleoproteins) by centrifugation at 75,000 × *g* for 1 h, the supernatant is dialyzed against 9.7 volumes of water, equivalent to 0.14 M sodium chloride. Centrifugation at 5,000 × *g* for 10 min gives a residue containing deoxyribonucleoproteins, but repeated dialysis of the residue in 1.5 M saline against water, as described above, proved to be better. The residual pellet is then extracted with 0.5 N sulfuric acid and trichloroacetic acid is added to the acid extract until a concentration of 20 % is attained, when basic proteins precipitate. The basic protein fraction is fully converted to the sulfate salt and then subjected to chromatography on Sephadex G-50 (eluent: 0.1 N acetic acid) which gives peaks of histones and a peak of protamine separately. When the specimen is obtained in November (in near maturity), only a small peak of lysine-rich histone remains, together with a large peak of protamine, in the chromatographic pattern. The basic protein fraction from the testis in full maturity (February and March), gives, of course, only a large peak of protamine, identified as pure iridine by analysis of amino acids and N-terminals, as well as by disc electrophoresis on acrylamide gel (0.1 M glycine-formic acid, pH 4.0, 3mA, 2.5 h).

d) DIXON and his colleagues, studying the biosynthesis of protamine in trout testis, have isolated and characterized protamine according to the following procedures:

(α) The homogenate of whole maturing testis cells of steelhead trout *(Salmo gairdnerii)*, in which spermatogenesis had been induced by injections of salmon pituitary extracts, is extracted with 0.2 N HCl to obtain basic proteins. The acid extract is then subjected to successive chromatography on Sephadex G-10 (40 × 2 cm), CM-Sephadex (H^+ form, elution with 0.1 N HCl), Sephadex G-25 (2.5 × 50 cm, 0.1 M acetic acid) and Bio-Gel P-10 (2 × 30 cm, 0.2 M acetic acid) to isolate and characterize a protamine (separated from histones) present in the extract (INGLES and DIXON, 1967). In the last chromatography on Bio-Gel P-10, patterns of whole histone and protamine are obtained reproducibly with maturity stages like those obtained by KOGA *et al.* (1969).

(β) More recently, DIXON's group extracted basic proteins with 0.2 N sulfuric acid from the homogenate of maturing testis cells or from purified testis chromatin prepared by the method of MARUSHIGE and BONNER (1966). The basic proteins are precipitated from the acid extract by the addition of 3—4 volumes of ethanol. After several hours of standing at −20 °C, the precipitate is collected by centrifugation. The precipitate is washed with ethanol, dissolved in water, and applied to a CM-cellulose column (1 × 7 cm, H^+ form). After washing the column extensively with water, the basic proteins are eluted with 0.2 N HCl. The eluate is lyophilized to leave nuclear proteins which are further chromatographed on a Bio-Gel P-10 column, as described above, to separate protamine from total histone. The pooled material from the effluent containing protamine is finally subjected to chromatography on a CM-cellulose column (1 × 30 cm) prepared according to SCHLOSSMAN *et al.* (1965). Application of a linear gradient of lithium chloride concentration (0.2 M:2.0 M LiCl, 350 ml:350 ml, both buffered with a 0.01 M lithium acetate-acetic acid buffer of pH 5.0) enables whole protamine to be eluted from the column satisfactorily (LING *et al.*, 1969. See also MARUSHIGE and DIXON, 1969, and MARUSHIGE *et al.*, 1969).

3. Some Remarks on the Methods of Preparation of Whole Protamines

It is apparent that there are many procedures for preparing whole protamines. With typical monoprotamines such as clupeine and salmine from mature testes or sperm, a combination of the procedures described in Sections A. 2 and C. 2. c) α or C. 2. c) β will give a high yield suitable for routine use. Although it may serve as a pure specimen of protamine, it usually consists of a mixture of several components. For special purposes which require the use of homogeneous components, such a specimen must be subjected to further fractionation (see Chap. VII).

For the investigation of the changes which occur in the contents of protamine as well as other nuclear basic proteins (histones) in fish testis during spermatogenesis, however, the methods developed by Dixon's group (C. 2. d) α or β) or by the Tokyo group (C. 2. c) γ) will give satisfactory results.

In future studies, the applicability of the procedures for preparing protamine in general should be carefully examined in each case because, of the very many kinds of protamine reported in the literature, only a few have been thoroughly studied; these are clupeine and protamines from Salmonidae, a group of strongly basic monoprotamines, and sturine, a triprotamine. On most of the remaining protamines we have but little information. We must also be careful when discussing the higher-order structure of protamines in general, as evidence is accumulating to suggest that some kinds of histone prepared by procedures involving the use of mineral acids, differ in structure from "native" histones in isolated chromatin (OHBA, Y., private communication). This may not be the case with the monoprotamines, including clupeine, salmine, and iridine, which have been shown to have only a random-coil structure in aqueous solution (see Chap. IX), but we know nothing at all about the primary and higher-order structure of many other protamines likely to be prepared and studied in future.

Chapter IV

Composition

A. Composition of Nucleoprotamines

1. Composition of Sperm Head or Sperm Nucleus

The dried sperm heads or sperm nuclei of herring, salmon, and trout, prepared according to the procedures described in Sections A. 1 and A. 2 of the last chapter, contain approximately 20% nitrogen, 6% phosphorus, and 30 to 33% arginine. From these values it is calculated that the sperm heads or sperm nuclei contain approximately 60 to 64% deoxyribonucleic acid and 40 to 36% protamine by weight, to give an approximately 1:1 molar ratio of phosphorus to arginine [FELIX *et al.*, 1951 (1, 2); FELIX and KREKELS, 1953; ANDO and HASHIMOTO, 1958 (1); and FELIX and HASHIMOTO, 1963]. The result suggests that in these nucleoprotamines DNA and protamine are held together by an electrostatic bond between the phosphate anions of the nucleic acid moiety and the guanidinium cations of arginine residues in the protamine moiety (FELIX, 1953). It has been observed, however, that during the early stage of spermatogenesis the immature testis cell nucleus fraction is low in phosphorus and arginine and that the content of both increases with maturity to attain the final values described above [ANDO and HASHIMOTO, 1958 (2); FELIX *et al.*, 1958].

2. Composition of Reconstituted Nucleoprotamine

The fibrous precipitate of nucleoprotamine (prepared and purified as described in Chap. III. B, by pouring the 2 M sodium chloride extract in dissociated form of sperm heads or sperm nuclei into water until the concentration of saline is 0.14 M) represents "reconstituted" or "recombined" nucleoprotamine. The composition of nucleoprotamine thus obtained is not necessarily identical to that in the original sperm heads or sperm nuclei. However, many workers have reported that the composition (the content of nitrogen, phosphorus, and arginine, and the molar ratios N/P, P/arginine, and DNA/protamine) of such reconstituted nucleoprotamine is fairly similar to that in the original material [POLLISTER and MIRSKY, 1946; WATANABE and SUZUKI, 1951 (1); SUZUKI and WATANABE, 1952 (1); FELIX *et al.*, 1951 (1, 2); FELIX *et al.*, 1952 (1)]. Table IV-1 gives typical examples of the composition of sperm heads, sperm nuclei, and reconstituted nucleoprotamines from various sources.

3. Composition of Artificial Nucleoprotamines

If a solution of protamine (for example, clupeine or salmine) is mixed with nucleic acid (deoxyribonucleic acid or ribonucleic acid), artificial nucleoprotamine preparations are obtained. By controlling the experimental conditions of mixing, it is possible to change

Table IV-1. Composition of sperm heads, sperm nuclei, and nucleoprotamines[a]

Composition Sources and Materials	P (%)	N (%)	Arginine (%)	DNA (%)	Protamine (%)	N/P (w/w)	P/Arginine (atom/ molecule)	Ref.
Salmo salar (Rhine salmon)								
Defatted sperm heads (A)				60.50	35.56			1
Sperm heads (A)				59.83	35.32			2
Nucleoprotamine	5.39	20.25	28.90	59.8[b]		3.57	1:0.96	3
Oncorhynchus keta (Hokkaido salmon)								
Nucleoprotamine	6.3	20.3		70		3.2		4
Salmo irideus (Rainbow trout)								
Sperm heads (B)	5.6	20.6	33.0			3.6	1:1.0	5
Cell nuclei from immature testis	3.3	16.4	12.1			5.0	1:0.7	5
Sperm nuclei	5.87	19.67	(30.86)[c]	62.7[b]		3.35	(1:0.94)[c]	6
Sperm nuclei	5.6	19.8	29.8			3.53	1:0.94	7
Nucleoprotamine	5.65	19.52	30.6	60.3[b]		3.43	1:0.96	6
Nucleoprotamine	6.6	20.2				3.1		4
Salmo fontinalis (Bachsaibling)								
Sperm nuclei	5.88	19.78	30.44	61.7[b]		3.36	1:0.92	8
Nucleoprotamine	5.73	19.67	30.15	60.1[b]		3.43	1:0.94	8
Salmo trutta (Bachforelle)								
Sperm nuclei	5.72	19.80	30.42	60.2[b]		3.46	1:0.95	8
Nucleoprotamine	5.71	19.68	30.20	60.1[b]		3.44	1:0.95	8
Clupea harengus (North Sea herring)								
Sperm heads (A)	3.93	17.72		39.4[b]		4.50		6
Sperm nuclei	5.92	19.72	30.42			3.42	1:0.94	9
Sperm nuclei	5.6	19.6	33.8			3.50	1:0.93	7
Nucleoprotamine	5.68	19.57	30.77	57.0[b]	37.0[b]	3.44	1:0.96	6
Clupea pallasii (Pacific herring)								
Nucleoprotamine	5.4	20.6		63		3.8		4
Acipenser sturio (German sturgeon, Baltic Sea)								
Nucleoprotamine	5.47	19.72	23.51	62.7[b]		3.60	1:0.76	3

[a] Sperm heads (A) and (B), sperm nuclei, and nucleoprotamines (reconstituted) are specimens prepared according to the procedures described in Chap. III. A. 1—3 and B., respectively.
[b] Values computed by the present authors using the values for P and arginine content given in the literature.
[c] Added afterwards in the author's review (FELIX, 1960), though not found in the original literature.

References to Table IV-1:

1. MIESCHER, F.: Die histochemischen und physiologischen Arbeiten, Bd. I und II. Leipzig 1897.
2. BURIAN, R.: Ergebn. Physiol. **5**, 806 (1906).
3. FELIX, K., FISCHER, H., KREKELS, A.: Z. physiol. Chem. **289**, 127—131 (1952).
4. SUZUKI, K.: Rp. Inst. Sci. and Technol. Univ. Tokyo **11** (No. 4), 177—217 (1957).
5. ANDO, T., HASHIMOTO, C.: J. Biochem. (Tokyo) **45**, 529—540 (1958). Values for specimens prepared from testes of rainbow trout obtained from the fish hatchery at Okutama near Tokyo in February and September, respectively.
6. FELIX, K., FISCHER, H., KREKELS, A., MOHR, R.: Z. physiol. Chem. **287**, 224—234 (1951).
7. FELIX, K., HASHIMOTO, C.: Z. physiol. Chem. **330**, 205—211 (1963).
8. FELIX, K., FISCHER, H., KREKELS, A., MOHR, R.: Z. physiol. Chem. **289**, 10—19 (1951).
9. FELIX, K., KREKELS, A.: Z. physiol. Chem. **293**, 284—286 (1953).

the composition of the artificial nucleoprotamines within certain ranges [MIESCHER, 1897; STEUDEL, 1913; STEUDEL and PEISER, 1922; SUZUKI, 1950; WATANABE and SUZUKI, 1951 (3); SUZUKI and WATANABE, 1952 (2), 1953; ALEXANDER, 1953; INOUE and ANDO, 1966].

Table IV-2. Base composition of deoxyribonucleates prepared from sperm nuclei of herring and salmonoids

Species	Adenine	Thymine	Guanine	Cytosine	5-Methyl-cytosine	Ref.
Salmo irideus	28	26	21	20	+	1
Salmo trutta	28	26	21	20		1
Salmo fontinalis	27	26	20	20	+	1
Salmo salar	27	26	21	21		1
Clupea harengus	27	26	21	19	2	1
Acipenser sturio	29	27	22	20		1
Clupea pallasii	29	29	21	22	3	2
Oncorhynchus keta	27	28	22	21	2	2
Salmo irideus	28.9	26.8	21.8	21.0	1.5	3
	29.5	26.1	22.4	21.4	0.6	
	29.7	26.8	21.2	21.6	0.7	
Oncorhynchus gorbuscha	28.5	27.0	22.5	20.4	1.7	4
Oncorhynchus keta	27.2	27.0	22.9	21.0	2.0	4
Oncorhynchus kisutch	28.2	27.0	22.8	20.9	1.4	4
Oncorhynchus nerka	27.3	27.4	22.9	20.4	1.8	4
Oncorhynchus tschawytscha	26.9	26.0	23.7	21.0	1.6	4
Salmo salar	27.8	26.5	23.6	20.6	1.6	4
Salmo gairdnerii	27.8	27.4	22.5	20.0	2.3	4
Salvelinus fontinalis	28.0	27.3	22.9	20.2	1.6	4

1. FELIX, K., JILKE, I., ZAHN, R. K.: Z. physiol. Chem. **303**, 140—152 (1956).
2. SUZUKI, K.: Rp. Inst. Sci. and Technol., Univ. Tokyo **11** (No. 14), 177—217 (1957).
3. KAWADE, Y.: Rp. Inst. Sci. and Technol., Univ. Tokyo **10** (No. 12), 149—196 (1956).
4. DIXON, G. H., SMITH, M.: Progress in nucleic acid research and molecular biology (DAVIDSON, J. N., COHN, W. E., Eds.). **8**, 9—34 (1968).

Refs. 1—3: Base composition in molar concentration per 100 atoms of phosphorus.
Ref. 4: Base composition in mole percent.

4. Base Composition of Deoxyribonucleic Acid Associated with Protamine

Data on the base composition of the deoxyribonucleates from the sperm nuclei of herring, salmon, and rainbow trout are summarized in Table IV-2. The values for salmonoids, described by DIXON and SMITH (1968) in their review, are also included for comparison.

B. Composition of Unfractionated (or Whole) Protamines

Protamines usually contain only a few kinds of amino acid. Their abnormally high content of basic amino acids, particularly arginine, is reflected in their extremely high nitrogen content, which often exceeds 30%. Every protamine contains arginine, but the kinds and amounts of monoamino acids are limited: alanine and serine appear to occur in all known protamines, proline and valine in most of them, glycine and isoleucine in many, and threonine in some. Only a few kinds of protamine contain acidic amino acids. In the protamines analyzed thus far, aromatic amino acids have been found in thynnine from tunny (tyrosine and a trace amount of phenylalanine), cyclopterine from *Cyclopterus lumpus* (tyrosine), galline from domestic fowl (tyrosine), and a basic protamine-like protein from bull (tyrosine and phenylalanine). The only sulfur-containing amino acids so far found in protamines are methionine in mugiline β, and six half-cystine residues in a protamine-like protein from bull sperm (see Chap. VII. B. 7).

1. Classification of Protamines

KOSSEL (1928) classified the protamines into three groups according to the number of kinds of basic amino acid each contained (cf. Table II-1 for classification of main protamines).

(i) *Monoprotamines*, which contain only one kind of basic amino acid: arginine. Most protamines belong to this group. Clupeine, salmine, iridine, and mugiline β are examples.

(ii) *Diprotamines*, which contain two kinds of basic amino acids: arginine plus either lysine or histidine. Cyprinine, crenilabrine, and barbine are reported to belong to the group containing lysine and arginine, and percine and lacustrine to that containing histidine and arginine.

(iii) *Triprotamines*, which contain all three basic amino acids. Sturine and some protamines are reported to belong to this group.

2. Ratio of Carbon to Nitrogen in Protamine

MIESCHER gave the chemical formula $C_9H_{21}N_5O_3$ for the protamine isolated as a platinum salt from Rhine salmon in 1874. In the same year PICCARD (1874) repeated MIESCHER's experiment and gave the formula $C_{16}H_{32}N_9O_2 \cdot 2HCl \cdot PtCl_4$ for the platinum salt. KOSSEL preferred the formula $C_{16}H_{31}N_9O_3 \cdot H_2SO_4$ for his specimen of salmine sulfate (KOSSEL 1896/1897). At first sight, such formulas seem to be of merely historical interest and to tell little about the chemical nature of protamine. Our Tokyo group (ANDO *et al.*, 1955), however, thought that the values for the carbon/nitrogen ratio (by weight) calculated on the basis of these formulas might be useful in characterizing the protamines, because an old work on protamines (GOTO, 1902) reminded them of the use of the ratio. The C/N ratio in the protamine was calculated from the formulas of MIESCHER, PICCARD, and KOSSEL and found to be 1.54, 1.49, and 1.49, respectively. Further, the C/N ratio was calculated from the data of older elementary analyses and recent amino-acid composition analyses for various kinds of protamines. The calculated values for protamines are usually within the range 1.5 to 2. These are extremely low in comparison with the values calculated for the C/N ratio in typical proteins. In the case of histones, which are present in the cell nuclei of higher organisms, also in association with DNA, the C/N ratio ranges from 2.4 to 2.9. Among amino acids, arginine gives the lowest C/N ratio, 1.29; histidine and glycine give a ratio of 1.71; lysine, alanine, and serine, 2.57; threonine, 3.43; valine and proline, 4.28; isoleucine or leucine, 5.15; and tyrosine and phenylalanine give the highest C/N ratio, 7.70. The extremely low values for the C/N ratio found with protamines (approximately 2 or less) may be regarded as a reflection of the singular amino-acid composition of these basic proteins.

3. Amino-Acid Composition

Though protamines have been studied for nearly a century, accurate and detailed studies on their amino-acid composition have been made only during the last two decades, and using limited numbers of a relatively few kinds of fish. Table IV-3 summarizes the results of recent analyses with unfractionated or partially fractionated protamine preparations, including protamine-like basic proteins isolated from sperm nuclei of fowl (galline) and bull. As is obvious from the table, comparative biochemical studies of amino-acid composition have been made in considerable number only for the salmonoid species. Some apparent difference in amino-acid composition is often found among specimens of protamines supposed to have been prepared from fish of the same species (*e.g.*, clupeine and salmonoid protamines). This kind of discrepancy is especially remarkable in the smaller amino-acid fractions. Deviations of this kind may be accounted for by differences in the maturity and freshness of the starting materials and methods of isolation and purification. The heterogeneous nature of protamines has been fully established in recent years (see Chap. VII), so that careless handling of protamines during preparation may well lead to an unexpected fractionation.

The amino-acid composition of whole thynnine was reported to change with the age of the fish, becoming much simpler in older individuals than in younger ones. Thus, whole thynnine from 5-year-old fish contained only seven kinds of amino acids (Arg, Val, Ala, Ser, Pro, Tyr and Glu), whereas that from 2 to 3-year-old fish had small or negligible amounts of seven more kinds (Thr, Lys, Gly, Asp, Phe, Leu and Ile) in addition to those found in the older fish (WALDSCHMIDT-LEITZ and GUTERMANN, 1966).

The amino acid composition of homogeneous components of some protamines is described in Chaps. VII and VIII.

Table IV-3. Amino-acid composition of some protamines

Protamine	Species (Name or origin of fish)	Ex-pres-sion	Gly	Ala	Val	Ile	Ser	Thr	Pro	Arg	His	Lys	Glu	Asp	Total	Notes	Ref.
Protamines of herring family																	
Clupeine	*Clupea harengus*	A[a]		1.89	1.60	0.43	2.42	0.65	2.22	89.7					98.91 %		1
	(Norwegian Sea and North Sea herring)	A[a]	0.43	2.52	1.39	0.35	2.24	1.00	2.69	89.6					100.2 %	As sulfate salt: total N, 28.4 %; H_2SO_4, 20.5 %	2
	Clupea harengus	A[a]		3.7	1.5	1.1	1.8	1.8	3.9	86.9					100.7 %		3
	Clupea harengus (Norwegian winter herring)	A[b]	0.3	2.4	1.4	0.3	2.6	1.0	1.9	89.0					98.9 %	(NH_3-N, 0.9 %)	4
	Clupea pallasii (Pacific herring)	A[a]	0.24	2.5	1.8	0.27	2.3	1.0	2.9	90.1					101 %	II; As sulfate salt: total N, 23.6 %; H_2SO_4, 20.5 %	5
		A[a]	0.36	2.6	1.7	0.45	2.4	1.3	2.9	91.0					103 %	II-1; total N, 24.4 %	5
		A[a]	0.48	2.3	1.1	0.60	2.4	1.4	2.4	90.3					101 %	II-2; total N, 23.8 %	5
		A[a]	0.57	2.7	1.2	0.57	2.5	1.0	2.5	89.0					100.1 %	Clup. I	6
		A[a]	0.22	2.5	1.7	0.27	2.6	0.97	2.5	88.0					98.8 %	Clup. II	6
		C[a]	1.3	7.4	4.4	0.8	7.9	3.4	7.3	65.5					100.0 %	Whole clup. sulfate. Orn, 2.0% (possibly derived from Arg).	6a
		C[a]	1.5	6.3	3.5	1.1	7.4	3.5	8.0	68.7					100.0 %	Whole clup. hydrochloride. Orn, trace.	6a
										87.3					110.1 g		7
		E[a]	0	4.7	3.6	1.0	3.4	1.9	8.2	86.8					109.6 g		7
Protamines of salmon family																	
Salmine	*Salmo salar* (Rhine salmon)	B[a]	3	2	4	—	5	—	5	55					(74)[c]		8

Table IV-3 (continued)

Protamine	Species (Name or origin of fish)	Ex-pres-sion	Gly	Ala	Val	Ile	Ser	Thr	Pro	Arg	His	Lys	Glu	Asp	Total	Notes	Ref.
Salmine	*Oncorhynchus tscha-wytscha*	A[b]	1.96	0.56	1.42	0.38	3.04		3.13	87.7					98.2 %		9
	(Chinook salmon;	B[b]	2.0	0.6	1.4	0.4	3.1		3.2	22.3					(33.0)[c]		9
	Columbia River in North America)	B[a]	2.2	0.6	1.7	0.4	3.4		2.6	22.1					(33.0)[c]		10
	Oncorhynchus kisutch (Coho salmon)	B[a]	2.2	0.6	1.8	0.3	3.6		2.7	21.8					(33.0)[c]		10
	Oncorhynchus keta	A[a]	2.1	0.32	1.6	0.34	3.3		2.9	89.8					99.71 %		6
	(Hokkaido salmon; Ishikari River in Hokkaido, Japan)	C[a]	6.5	1.3	4.7	1.3	9,9	0	9.1	67.2					100.0 %		11
	Commercial preparation from Columbia	A[a]		1.8	1.6	0.5	3.0		3.0	90.2	0	0	0	0	100.1 %	As sulfate salt: H_2SO_4, 19.1 %	12
	River salmon	E[a]		3.6	4.1	1.5	7.0		7.9	88.4					112.5 g		12
	Unknown	A[a]	1.78	0.56	1.20	0.56	3.94		2.29	89.0					99.33 %	As sulfate salt: H_2SO_4, 19.85 %	13
		E[a]	2.94	1.10	3.1	1.61	9.1		5.8	85.2					108.85 g		13
	Unknown	A[b]	2.17	0.30	1.47	0.62	2.41		1.96	91.0					98.26 %		14
		E[a]	3.6	0.6	3.8	1.8	5.6		5.0	87.5					107.9 g		14
	(Columbia River salmon?)	A[b]	1.96	0.75	1.56	0.41	2.96		3.06-3.33	90.4					101.10-101.37 %		7
	Oncorhynchus Genus	A[a]	1.97	0.59	1.85		2.70		2.16	90.6					99.87 %		15
	(Caught near Vancouver, British Columbia)	A[a]	1.90	0.55	1.81	—	2.70		2.70	86.3					95.69 %		16

Table IV-3 (continued)

Protamine	Species (Name or origin of fish)	Expression	Gly	Ala	Val	Ile	Ser	Thr	Pro	Arg	His	Lys	Glu	Asp	Total	Notes	Ref.
Salmine	*Oncorhynchus Genus* (Preparation from Boots Pure Drug Co., Ltd.)	A[a]	1.80	0.45	1.40	0.44	3.12		2.70	89.8					99.71 %		17
Protamines of trout family																	
Iridine	*Salmo irideus* (Rainbow trout;	A[a]	1.9	0.6	1.7	0.3	3.3		2.8	89.6					100.2 %	As sulfate salt: total N, 23.0 %	18
	Okutama hatchery near Tokyo)	C[a]	6.2	2.2	5.4	1.0	9.6	0	7.1	68.6					100.1 %		11
	(Rainbow trout, Regenbogenforelle)	A[a]	2.3	0.6	1.8	0.3	2.7	0	2.9	89.6	0	0			101.0 %	NH_3-N, 0.8 %	19
	(Rainbow trout, Regenbogenforelle)	B[a]	2	2	4	1	3	—	5	50	—	2	—	—	(69)[c]	As sulfate salt: total N, 4.07 %	20
		A[a]	1.93	0.57	1.71	0.26	2.84	—	3.37	89.2					99.9 %	As sulfate salt: total N, 26.8 %; SO_4^{--}, 20.1 %	2
	Salmo gairdnerii (Rainbow or steel-head trout)	B[a]	2.2	0.5	1.8	0.3	3.6		2.7	22.1					(33.0)[c]		10
Truttin	*Salmo trutta forma fário* (Bachforelle)	A[a]	2.2	0.5	2.0	0	2.7	0	3.0	90.5	0	0			101.9 %	NH_3-N, 0.8 %	19
Lacustrin	*Salmo trutta forma lacustris* (Seeforelle)	A[a]	2.2	0.5	2.0	0	2.7	0	3.0	90.2	0	0			101.9 %	NH_3-N, 0.8 %	19

Table IV-3 (continued)

Protamine	Species (Name or origin of fish)	Expression	Gly	Ala	Val	Ile	Ser	Thr	Pro	Arg	His	Lys	Glu	Asp	Total	Notes	Ref.
Protamines of Whitefish family from Lake Constance																	
Coregonine	*Coregonus lavaretus nat. bodanensis oekot. pelagicus* (Blaufelchen)	A[a]	2.4	0.4	2.1	0	2.9	0.4	3.6	88.9	0	0			101.3%	NH_3-N, 0.6%	19
	Coregonus lavaretus nat. bodanensis oekot. litoralis (Bodenrenke)	A[a]	2.1	0.3	2.1	0	2.8	0.4	3.4	88.6	0	0			100.7%	NH_3-N, 1.0%	19
	Coregonus lavaretus nat. bodanensis oekot. nanus (Gangfisch)	A[a]	2.3	0.2	2.3	0	3.1	0.4	3.5	89.0	0	0			101.6%		19
Protamines of char family																	
Fontinine	*Salvelinus fontinalis* (Canadian char; Bachsaibling)	A[a]	2.1	0.2	2.1	0	2.9	0	2.9	89.9	0	0			100.7%	NH_3-N, 0.6%	19
	Salvelinus alpinus (Schwarzreuther, from the Königssee; Seesaibling)	A[a]	2.0	0.3	2.0	0	3.1	0	3.0	89.9	0	0			101.2%	NH_3-N, 0.9%	19
Other protamines																	
Mugiline β	*Mugil japonicus*	A[a]		2.08	0.92	1.28	1.06	1.02	3.21	87.71			1.08	0.33	99.98%	Met, 0.14%; NH_3-N, 1.15%	21
		A[a]		1.98	1.05	1.28	1.01	1.06	2.52	89.68			0.82	—	100.04%	No Met; NH_3-N, 0.64%	21
Thynnine	*Thynnus thynnus*	A[a]	0.4	2.4	2.0	0.1	2.4	0.9	2.2	85.2		0.8	0.7	0.1	99.4	Leu, 0.1%; Tyr, 0.9%; Phe, 0.1%; NH_3-N, 1.1%	22
Sturine	*Acipenser sturio* (Baltic sturgeon)	B[a]	2	5	—	2	3	1	—	35	7	9	1		(65)[c]		8

Table IV-3 (continued)

Protamine	Species (Name or origin of fish)	Ex-pres-sion	Gly	Ala	Val	Ile	Ser	Thr	Pro	Arg	His	Lys	Glu	Asp	Total	Notes	Ref.
Stelline	*Acipenser stellatus*	C[a]	4.4	4.6		2.9	5.4	2.3	1.2	56.2	8.4	12.9	1.7		100.0%		23
Galline	(Fowl)	D[a]	2.80	1.22	0.68		3.89	0.72	1.94	19.1	0.51	0.0	0.45	0.28	92.0	(Leu + Ile + Phe), 0.35%; Met, 0.0;% Tyr, 1.58%; NH_3-N, 1.17%; total-N, 24.4%	24
	Gallus domesticus	A[a]								88.0							25
		B[b]	1	5	3	1	5	2	5	42[b]	—		1	—	(65)[c]	As HCl salt: total N, 25.6%, no Trp and Tyr	25
	Gallus domesticus	C[a]	8.7	3.5	2.2		16.9	1.6	3.5	55.4	0.3	0.9	0.7	0.5	100.0%	Leu, 0.2%; Tyr, 5.8%	26
A basic protein from bull sperm heads		B[a]	2.1	1.0	2.0	1.0	1.9	3.0	0	24.3	1.0	0	1.0	0	(47.2)[c]	Purified Fract. II: Aminoethyl-Cys, 5.9%; Tyr, 2.0%; Leu, 1.0%; Phe, 1.0%; Met, 0%	27

[a] Values described in the original literature.
[b] Values computed by the present authors from those in original literature.
[c] Values somewhat arbitrarily chosen by the author(s).

A Amino-acid N as a percentage of total-N.
B Molar ratios of constituent amino-acids.
C Mole percent of constituent amino-acids.
D Moles amino-acid per 100 moles N.
E Grams of amino-acid per 100 grams of protamine.

Abbreviations used after the references:
AAA analyzed using automated amino-acid analyzer according to SPACKMAN, D. H., STEIN, W. H., MOORE, S.: Anal. Chem. **30**, 1190—1206 (1958).
IEC ion exchange chromatography according to the procedure of MOORE, S., STEIN, W. H.: J. biol. Chem. **192**, 663—681 (1951).
PC partition chromatography on paper.
SC partition chromatography on starch gel column according to the procedure of STEIN, W. H., MOORE, S.: J. biol. Chem. **176**, 337—365, 367—388 (1948), and MOORE, S., STEIN, W. H.: J. biol. Chem. **178**, 53—77 (1949).

References to Table IV-3:

1. FELIX, K., FISCHER, H., KREKELS, A., RAUEN, H. M.: Z. physiol. Chem. **286**, 67—78 (1950). PC.
2. FELIX, K., HASHIMOTO, C.: Z. physiol. Chem. **330**, 205—211 (1963). AAA.
3. KURODA, Y.: J. Biochem. (Tokyo) **38**, 115—118 (1951). PC-copper method.
4. WALDSCHMIDT-LEITZ, E., GUDERNATSCH, H.: Z. physiol. Chem. **309**, 266—275 (1957). SC. Average (computed by the present authors) of values of three specimens (III, IV, and V in their Table II).
5. YAMASAKI, M.: Sci. Pap. Coll. Gen. Educ., Univ. Tokyo **8**, 165—173 (1958). Determined as 2,4-dinitrophenyl derivatives. II—1, II—2, and II—3, the less, more, and most soluble fractions, respectively, fractionated by utilizing the temperature dependence of the solubility of the picrate salt in 67% acetone-water.
6. ANDO, T., ISHII, S., SATO, M.: J. Biochem. (Tokyo) **46**, 933—940 (1959). IEC, together with the method of S. ISHII: J. Biochem. (Tokyo) **43**, 531—537 (1956). Clup. I and II indicate specimens prepared from mature milts of herring caught off the NE coast of Hokkaido (Mombetsu, facing the Sea of Okhotsk, May 7, 1947) and off the W coast of Hokkaido (Yoichi, facing the Sea of Japan, April 3, 1951), respectively.

6 a. KAWASHIMA, S.: Unpublished data (1969).

7. BLOCK, R. J., BOLLING, D., GERSHON, H., SOBER, H. A.: Proc. exp. Biol. (N. Y.) **70**, 494—496 (1949). Combination of chemical and microbiological techniques.
8. FELIX, K., FISCHER, H., KREKELS, A.: Z. physiol. Chem. **289**, 127—131 (1952). PC.
9. CALLANAN, M. J., CARROLL, W. R., MITCHELL, E. R.: J. biol. Chem. **229**, 279—287 (1957). IEC.
10. INGLES, C. J.: Ph. D. Thesis, Univ. of Brit. Columbia, Vancouver, Canada, 1968; cited in DIXON, G. H., SMITH, M.: Nucleic acids and protamine in Salmon testes. In: Progress in nucleic acid research and molecular biology (DAVIDSON, J. N., COHN, W. E., Eds.) **8**, pp. 9—34 (1968).
11. WATANABE, S.: Dr. Thesis, Univ. Tokyo, June 1969. AAA.
12. BLOCK, R. J., BOLLING, D.: Arch. Biochem. Biophys. **45**, 419—424 (1945). Combination of chemical and microbiological techniques.
13. TRISTRAM, G. R.: Nature (Lond.) **160**, 637 (1947).
14. HAMER, D., WOODHOUSE, D. L.: Nature (Lond.) **163**, 689—690 (1949). PC.
15. MILLS, G. L.: Biochem. J. **50**, 707—712 (1952). Determined as 2,4-dinitrophenyl amino acid derivatives.
16. VELICK, S. F., UDENFRIEND, S.: J. biol. Chem. **191**, 233—238 (1951). Isotope derivative method, ion-exchange chromatography on Dowex-50 column, PC, and microbiological assay.
17. CORFIELD, M. C., ROBSON, A.: Biochem. J. **55**, 517—522 (1953). SC.
18. ANDO, T., HASHIMOTO, C.: J. Biochem. (Tokyo) **45**, 529—540 (1958). IEC., together with the method of ISHII, S.: J. Biochem. (Tokyo) **43**, 531—537 (1956).
19. WALDSCHMIDT-LEITZ, E., GUTERMANN, H.: Z. physiol. Chem. **323**, 98—104 (1961). An automated apparatus of K. HANNIG: Clin. chim. Acta **4**, 51—57 (1959).
20. FELIX, K., FISCHER, H., KREKELS, A., MOHR, R.: Z. physiol. Chem. **287**, 224—234 (1951). PC.
21. OTA, S., MURAMATU, M., HIROHATA, R., OKUDA, Y., YANG, C.-C., KAO, K.-C., CHIN, W.-C., CHANG, C.-C., IMAI, Y., ONO, T.: Ann. Rep. Lab. Protein Chem., Yamaguchi Medical School **1**, 1—10 (1966). Values for specimens of lot No. 3 in Table 5 and lot No. 9′ in Table 8 of their report, respectively. IEC.
22. WALDSCHMIDT-LEITZ, E., GUTERMANN, H.: Z. physiol. Chem. **344**, 50—54 (1966). An automated apparatus of K. HANNIG: Clin. chim. Acta **4**, 51—57 (1959).
23. RAUKAS, E.: Eesti NSV Teaduste Akad. Toimetised, Biol. Seer. **15**, 342—345 (1966).
24. DALY, M. M., MIRSKY, A. E., RIS, H.: J. gen. Physiol. **34**, 439—450 (1951). SC.
25. FISCHER, H., KREUZER, L.: Z. physiol. Chem. **293**, 176—182 (1953). PC.
26. NAKANO, M., TOBITA, T., ANDO, T.: Biochim. biophys. Acta (Amst.) **207**, 553 (1970). AAA.
27. COELINGH, J. P., ROZIJN, T. H., MONFOORT, C. H.: Biochim. biophys. Acta (Amst.) **188**, 353—356 (1969). COELINGH, J. P.: Dissertation, The State University, Utrecht, Oct. 1971. AAA.

Chapter V

Molecular Weight

A. Nucleoprotamines

The molecular weight of the DNA in fish sperm nucleus as determined by the sedimentation-viscosity method falls between 7 and 20 million, being 6.8 million for herring (KAWADE and WATANABE, 1956), 10.7 million for salmon (LITT, 1958), 16.6 million for salmon (GEIDUSCHEK, 1962; HAMAGUCHI and GEIDUSCHEK, 1962), and 19.5 million for salmon (LETT *et al.*, 1962). Since DNA-protamine ratio in fish sperm nucleus or nucleoprotamine is estimated to be approximately 60:40 (w/w), as described in the preceding chapter, the molecular weight of a nucleoprotamine such as nucleoclupeine or nucleosalmine in the nucleus is calculated to be between 12 and 33 million.

B. Protamines (Whole or Unfractionated)

It is characteristic of protamines that they have a molecular weight below 10,000. They can therefore usually pass through a cellophane membrane. A preliminary observation by the sedimentation-velocity method was once reported to indicate a molecular weight in the range 1,700 to 3,000 for both clupeine and salmine preparations (SVEDBERG, T. and B. SJÖGREN, cited in WALDSCHMIDT-LEITZ *et al.*, 1931). However, the calculation of molecular weight based on the primary structure (see Chap. VIII) of protamine components recently obtained in homogeneous form gave values of about 4,000 to 4,250 for both clupeine and salmine (ANDO *et al.*, 1962; ANDO and SUZUKI, 1966, 1967; ANDO and WATANABE, 1969).

During the intervening 40 years the estimation of the molecular weight of unfractionated or whole protamines has been carried out by various chemical and physical methods. Some of these values for typical protamines, clupeine and salmine, are listed in Table V-1 together with the methods used and literature references.

For several other protamines, the following molecular weights have been estimated: for iridine 5,000 (ANDO and HASHIMOTO, 1958; HASHIMOTO, 1958) and 5,500 (FELIX and HASHIMOTO, 1963), both obtained by UV-absorption analysis of the DNP derivative; also for iridine 4,100 by the sedimentation-diffusion method (GEHATIA and HASHIMOTO, 1963), and 4,307 and 4,320 for fractionated iridine components I a and I b, respectively, based on the primary structures (ANDO and WATANABE, 1969). Approximately 7,000 to 9,000 was estimated for mugiline β from *Mugil japonicus* TEMMINCK and SCHLEGEL by various methods including sedimentation-diffusion, light scattering, methoxyl determination, and amino-acid analysis (MORISAWA, 1957; HIROHATA, 1958; OTA *et al.*, 1959, 1966). A value of about 5,400 was found for whole thynnine from tunny *(Thynnus thynnus)* by the ultracentrifuge sedimentation method using double-layered cells (WALDSCHMIDT-LEITZ and GUTERMANN, 1966), and values of 4,770, 4,486, and 4,458 were obtained for the thynnine Y2, Z1 and Z2

Table V-1. The molecular weight of clupeine and salmine

Molecular weight		Method	Reference
Clupeine	Salmine		
Unfractionated (whole)			
4000—4100		Electrometric titration	1
4200—4600	4200—4500	Electrometric titration	2
4470		Amino-acid analysis	3
	7930	Amino-acid analysis	4
	9570—10,240	Amino-acid analysis	5
	7000	Amino-acid analysis and analysis of N^{α}-pipsyl derivative	6
4500		Formol titration of methylester·HCl	3
4650		Methoxyl determination of methyl ester	3, 7
6000	6000	Methoxyl determination of methyl ester	8
4100		Determination of CO_2 evolved by the reaction with N-bromosuccinimide[a]	9
5600—7100	6600—7000	UV absorption of N^{α}-DNP-derivative	10
	3800 (± 10%)	UV absorption of N^{α}-DNP-derivative	11
	4000	UV absorption of N^{α}-DNP-derivative	12
4800 (± 400)	4800 (± 400)	UV absorption of N^{α}-DNP-derivative	13
4200—4700	4000—4500	UV absorption of N^{α}-DNP-derivative	14
5500		UV absorption of N^{α}-DNP-derivative	15, 16
6500		Monomolecular film	17
5000—6000	5000—6000	Diffusion-viscosity	18
2000—10,000		Sedimentation-diffusion	19
5000	5000	Sedimentation-diffusion	20
Fractionated			
4110 (YI component)		Calculated from the primary structure determined	21
4047 (YII component)		Calculated from the primary structure determined	22
4163 (Z component)		Calculated from the primary structure determined	23
	4250 (AI component)	Calculated from the primary structure determined	24
	4236—4392 (AII component)	Amino-acid analysis	24

[a] According to the method of Chappelle, E. W., and Luck, J. M.: J. biol. Chem. **229**, 171 (1957).

References to Table V-1:

1. Rasmussen, K. E., Linderstrøm-Lang, K.: Z. physiol. Chem. **227**, 181 (1934); Linderstrøm-Lang, K.: Trans. Faraday Soc. **31**, 324 (1935).
2. Hashimoto, C.: Nippon Kagaku Zasshi (J. Chem. Soc. Japan, Pure Chem. Sect.) **80**, 800 (1959); Ando, T., Iwai, K., Yamasaki, M., Hashimoto, C., Kimura, M., Ishii, S., Tamura, T.: Bull. Chem. Soc. Japan **28**, 406 (1953).
3. Felix, K., Mager, A.: Z. physiol. Chem. **249**, 111 (1937).
4. Tristram, G. R.: Nature (Lond.) **160**, 637 (1947).
5. Corfield, M. C., Robson, A.: Biochem. J. **55**, 517 (1953).
6. Velick, S. F., Udenfriend, S.: J. biol. Chem. **191**, 233 (1951).
7. Felix, K., Fischer, H., Krekels, A., Rauen, H. M.: Z. physiol. Chem. **286**, 67 (1950).

References to Table V-1 (continued):

8. ANDO, T., IWAI, K., KIMURA, M.: J. Biochem. (Tokyo) **45**, 27 (1958); ANDO, T., ISHII, S., HASHIMOTO, C., YAMASAKI, M., IWAI, K.: Bull. Chem. Soc. Japan **25**, 132 (1952). Cf. also ANDO, T., IWAI, K., YAMASAKI, M., HASHIMOTO, C., KIMURA, M., ISHII, S., TAMURA, T.: Bull. Chem. Soc. Japan **26**, 406 (1953).
9. ZIMMERMANN, E.: Dissertation, Johann Wolfgang Goethe-Universität, Frankfurt a. M. 1959.
10. HASHIMOTO, C.: Bull. Chem. Soc. Japan **28**, 385 (1955).
11. PHILLIPS, D. M. P.: Biochem. J. **60**, 403 (1955). A commercial specimen of salmine was used.
12. CALLANAN, M. J., CARROL, W. R., MITCHELL, E. R.: J. biol. Chem. **229**, 279 (1957).
13. ANDO, T., YAMASAKI, M., ABUKUMAGAWA, E., ISHII, S., NAGAI, Y.: J. Biochem. (Tokyo) **45**, 429 (1958).
14. YAMASAKI, M.: Sci. Pap. Coll. Gen. Educ., Univ. Tokyo **9**, 31 (1959).
15. ŠORM, F., ŠORMOVA, Z.: Collect. Czech. Chem. Commun. **16**, 207 (1951).
16. FELIX, K., HASHIMOTO, C.: Z. physiol. Chem. **330**, 205 (1963).
17. IMAHORI, K.: Unpublished data.
18. ISO, K., KITAMURA, T., WATANABE, I.: Nippon Kagaku Zasshi (J. Chem. Soc. Japan, Pure Chem. Sect.) **75**, 342 (1954).
19. DAIMLER, B. H.: Kolloid Z. **127**, 97 (1952).
20. UI, N.: Rep. Inst. Sci. and Techn. Univ. Tokyo **8**, 255 (1954).
21. ANDO, T., SUZUKI, K.: Biochim. biophys. Acta (Amst.) **140**, 375 (1967).
22. ANDO, T., SUZUKI, K.: Biochim. biophys. Acta (Amst.) **121**, 427 (1966).
23. ANDO, T., IWAI, K., ISHII, S., AZEGAMI, M., NAKAHARA, C.: Biochim. biophys. Acta (Amst.) **56**, 628 (1962).
24. ANDO, T., WATANABE, S.: Int. J. Protein Res. **1**, 221 (1969).

components resulting from further fractionation, which appeared to be homogeneous (BRETZEL, 1971). For whole galline from domestic fowl *(Gallus domesticus)*, a value of 8,653 was calculated from the amino-acid composition data (FISCHER and KREUZER, 1953) and approximately 7,000 was obtained by the gel-filtration method (NAKANO *et al.*, 1968, 1969, 1970, 1972). According to a preliminary report by Dutch workers, a basic protein (fraction II), obtained and purified from bull sperm heads and regarded as a protamine-like histone, has a molecular weight of 6,200 per subunit as calculated from the results of amino-acid analysis (COELINGH *et al.*, 1969; COELINGH, 1971; see Chap. VII. B. 7).

Chapter VI

Chemical Structure of Nucleoprotamines and Protamines

A. Nucleoprotamines

As mentioned above, the DNA and protamine content in sperm nuclei of herring, salmon, char, rainbow trout, etc., as well as in the reconstituted nucleoprotamines, has been observed to be approximately 60% and 40% (w/w), respectively [MIESCHER, 1897; FELIX *et al.*, 1951 (1, 2); WATANABE and SUZUKI, 1951; FELIX *et al.*, 1952 (1)], and the molar ratio of phosphorus to arginine in the complex to be 1:0.92—1.07 [FELIX *et al.*, 1951 (1, 2); FELIX *et al.*, 1952 (1); FELIX and KREKELS, 1953; ANDO and HASHIMOTO, 1958 (1, 2, 3); FELIX and HASHIMOTO, 1963]. Accordingly, an average about 1,670 molecules of monoprotamine will combine with one molecule of DNA, forming salt-like linkages between the arginine residues in protamine and the phosphate groups in DNA, assuming the molecular weight of protamine to be 4,000 and that of DNA 10,000,000 (cf. Chap. V). The physical structure of such nucleoprotamine complexes is described in Chap. IX.

B. Protamines (Whole or Unfractionated)

Monoprotamines such as clupeine, salmine, and iridine have a very simple amino-acid composition and contain no sulfur-containing amino acids. They should therefore be expected to have one N-terminal amino or imino and one C-terminal carboxyl group. Although protamines were among the earliest proteins to be studied, their primary structures have only recently been determined. Protamine molecules contain enormously large numbers of arginine residues, so that separation of the strongly basic peptides obtained by partial hydrolysis is fairly difficult. Furthermore, the location of such peptides, each of which contains one or more arginine residues, in the original peptide chain is not easy to establish. Another difficulty is that protamines consist of a mixture of several rather similar species of molecules and their separation by fractination has been accomplished only recently for a few protamines (cf. Chap. VII). The results described in the present chapter refer only to unfractionated or whole protamines.

N- and C-terminal residues and a large number of peptides were obtained from protamines. However, it was not known at the time when they were obtained, which component of a protamine they were derived from.

Nevertheless, from these results a gross structure can be suggested for each protamine, since all the components of a protamine are of fairly similar structure.

1. Amino-Terminal Residue and Sequence

The use of the method of dinitrophenylation (DNP) (SANGER, 1945) enabled PORTER and SANGER (1948) to confirm the presence of proline at the N-terminus of salmine. Since then, many workers have studied N-terminal residues and sequences in several protamines, using DNP or other modern methods. Both proline and alanine were first found as N-terminal amino acids in clupeine from Pacific herring *(Clupea pallasii)*, clearly indicating its heterogeneous character [Ando *et al.*, 1957 (1, 2), 1958; Yamasaki, 1959 (1, 2)]. This heterogeneity was also inferred from the result of the action of DFP-treated[1] leucine aminopeptidase on clupeine (Ando *et al.*, 1957). The same N-terminal amino acids were later observed to be present also in clupeine from North Sea herring *(Clupea harengus)* (Felix and Hashimoto, 1963). These results are shown in Table VI-1.

Before these results were obtained, a good many data were accumulated by various classical methods on the presence or absence of N-terminal amino or imino groups in clupeine and salmine (Kossel and Cameron, 1912; Kossel and Gawrilow, 1912; Felix and Dirr, 1929; Felix and Mager, 1937; Felix *et al.*, 1950; Waldschmidt-Leitz *et al.*, 1931; Rasmussen and Linderstrøm-Lang, 1934; Linderstrøm-Lang, 1935; Tristram, 1949).

2. Carboxyl-Terminal Residue and Sequence

The presence of C-terminal arginine in several protamines was confirmed by means of modern techniques using DFP-treated carboxypeptidase B, hydrazinolysis, and thiohydantoin procedures [Ando *et al.*, 1958 (1); Tobita *et al.*, 1968; Akabori *et al.*, 1952, 1956; Kawanishi *et al.*, 1964; Tristram, 1947, 1949]. These results are also shown in Table VI-1.

These results explained reports of the presence (Rasmussen and Linderstrøm-Lang, 1934; Linderstrøm-Lang, 1935; Ando *et al.*, 1952; Hashimoto, 1959; Felix and Mager, 1937; Ando and Hashimoto, 1950; Rauen *et al.*, 1952) or absence (Fraenkel-Conrat and Olcott, 1947; Herriott *et al.*, 1946) of the C-terminal carboxyl group in protamine, which had puzzled workers in this field. The possible presence of arginine at the C-terminus of a protamine had earlier been suggested by the release of a small amount of urea by the action of arginase [Felix and Lang, 1930; Felix, 1931; Dirr and Felix, 1932 (1); Felix *et al.*, 1932]. This enzyme was known to require both free guanidino and carboxyl groups for its activity (Felix *et al.*, 1928; Felix and Schneider, 1938; Hellerman and Perkins, 1935/36). At about the same time, Waldschmidt-Leitz *et al.* (1931) observed the liberation of arginine from the C-terminal of clupeine and salmine by the action of a protease from hog pancreas, which they called protaminase, later identified as carboxypeptidase B (Weil *et al.*, 1959). The detection, by hydrazinolysis, of small amounts of alanine, valine, proline, and serine in addition to arginine as C-terminal amino acid residues of some protamines, as listed in Table VI-1, must be ascribed to artifacts.

3. Amino-Acid Sequences of Peptides Obtained by Partial Hydrolysis — Hypothetical Structures of Whole Protamines

Around 1900, Kossel (1898) found a degradation product, which he called clupeone, in the acid hydrolyzate of clupeine. He fractionated the product by adding alcohol to several fractions, each of which was found to have the same proportion arginine-N to total N as the original clupeine (88—89%) (Kossel and Pringle, 1906). From this result Kossel deduced that the constitutional ratio of arginine (A) to monoamino acid (M) was A_2:M in clupeine and that the ratio was retained unchanged in every one of its degradation fragments. Accordingly, he assumed there was a primary structure common to protamines such as clupeine and salmine, as follows:

H-AAM-AAM-AAM-··· or H-AMA-AMA-AMA-··· .

[1] DFP = diisopropylfluorophosphate or diisopropylphosphofluoridate.

Table VI-1. N- and C-terminal amino-acids of some protamines

Protamine	N-Terminal	Method and Reference	C-Terminal	Method and Reference
Salmine (Species unknown)	Pro	DNP [1] I^{131}-Pipsyl derivative [2] PTC [3] *o*-Benzoquinone derivative [4] Carbamino derivative [5]		
Salmine *(Oncorhynchus keta)*	Pro	DNP [6—8] UV absorption of DNP-derivative [6—9] Leucine aminopeptidase [10]	Arg	DFP-treated carboxypeptidases B and A digestion [34, 35]
	Imino group	Manometric Van Slyke method [7, 11] Electrometric titration [12]		Hydrazinolysis [36, 37] Tryptic hydrolysis of salmine methylester [35]
	Pro·Arg·Arg	Tryptic hydrolysis of DNP-derivative [13—15]	Arg_{3-4}	DFP-treated carboxypeptidase B [34, 35]
Salmine *(Salmo salar)*	Pro	DNP [16] PTC [17]	Arg	Arginase digestion [38—41] Pancreatic protaminase digestion [42]
			Arg Ala (minor) Val (minor) Pro (minor)	Hydrazinolysis [22]
			Arg	Hydrazinolysis [43]
Salmine *(Oncorhynchus)*	Pro Pro·Arg·Arg	DNP [18] Tryptic hydrolysis of DNP-derivative [18]		

Table VI-1 (continued)

Protamine	N-Terminal	Method and Reference	C-Terminal	Method and Reference
Clupeine *(Clupea harengus)*	Pro	DNP [16, 19, 20]	Arg	Arginase digestion [38—41]
	Ser (minor)	DNP [19, 20]		Pancreatic protaminase digestion [42]
	Pro·Ala	PTC [17, 21, 22] DNP [17]		Hydrolysis of benzoylacetyl derivative [40]
	Pro (42%) Ala (58%)	UV absorption of DNP-derivative [23]		NBS[a] [44] Hydrazinolysis [43]
	Pro (36%) Ala (58%)	UV absorption of DNP-derivative [24]	Arg Ala (minor) Val (minor) Pro (minor)	Hydrazinolysis [22]
Clupeine *(Clupea pallasii)*	Ala	UV absorption of DNP-derivative [6, 7, 9]	Arg	DFP-treated carboxypeptidases B and A digestion [34, 35]
	Ala Pro	DNP [8, 13] Leucine aminopeptidase [10]		Hydrazinolysis [36, 37] Tryptic hydrolysis of clupeine methylester [35]
	Pro (42%) Ala (58%)	UV absorption of DNP-derivative [24]		
	Amino and imino groups	Manometric Van Slyke method [7, 11]	Arg_{3-4}	DFP-treated carboxypeptidase B [34, 35]
	Ala·Arg·Arg Pro·Arg·Arg	Tryptic hydrolysis of DNP-derivative [13—15]		
Clupeine (Unstated species, probably *Clupea harengus*)	Pro	DNP [25 a, 25 b]	Arg	Carboxypeptidase digestion [45]
	Arg (minor)	DNP [26]		Thiohydantoin method [45]
	Pro Ala	DNP after partial fractionation [27]		

Table VI-1 (continued)

Protamine	N-Terminal	Method and Reference	C-Terminal	Method and Reference
Iridine *(Salmo irideus)*	Pro	DNP [16, 23, 28] PTC [28] UV absorption of DNP-derivative [29]	Arg	DFP-treated carboxypeptidase B digestion [28, 29] Hydrazinolysis [23, 36, 37, 43]
			Arg, Ala, Val	Hydrazinolysis [22]
Truttine *(Salmo trutta)*	Pro Pro·Val	DNP [16] PTC [17, 21, 22]	Ala, Val, Pro	Hydrazinolysis [22]
			Arg	Hydrazinolysis [43]
Fontinine *(Salmo fontinalis)*	Pro Pro·Val	DNP [16] PTC [17, 21, 22]	Ala, Val, Ser	Hydrazinolysis [22]
			Arg	Hydrazinolysis [43]
Mugiline βI[b] *(Mugil japonicus)*	Pro, Arg (?,minor) Pro·Arg	DNP [30, 31] PTC [30, 31] Tryptic hydrolysis of DNP-derivative [31]	Arg	Hydrazinolysis [30, 31]
Mugiline β *(Mugil japonicus)*	Pro	DNP [32]	Arg	Hydrazinolysis [32]
Sturine *(Acipenser sturio)*	Ala, Glu	DNP [16] PTC [17, 21, 22]	Arg, Ala, Val	Hydrazinolysis [22]
			Arg	Hydrazinolysis [43]
Thynnine *(Thynnus thynnus)*	Pro, Ser (minor), Ala (minor)	DNP [33]	Arg, Ser (minor)	DFP-treated carboxypeptidase [33]
Galline *(Gallus domesticus)*	Ser, Ala, Arg	DNP [46] DNS [46]	Arg, Tyr	DFP-treated carboxypeptidases B and A digestio [46] Hydrazinolysis [46]

[a] Agmatine and a trace of putrescine were detected in the reaction mixture with clupeine. By the action of N-bromosuccinimide (NBS) on a peptide, the C-terminal residue is reported to convert into an amine residue with evolution of CO_2 [cf. CHAPELLE, E. W., LUCK, J. M.: J. biol. Chem. **229**, 171 (1958)].

[b] A major fraction from partial fractionation of mugiline β by countercurrent distribution.

References to Table VI-1:

1. PORTER, R. R., SANGER, F.: Biochem. J. **42**, 287 (1948).
2. VELICK, S. F., UDENFRIEND, S.: J. biol. Chem. **191**, 233 (1951).
3. LANDMANN, W. A., DRAKE, M. P., DILLAHA, J.: J. Amer. chem. Soc. **75**, 3638 (1953).
4. MASON, H. S., PETERSON, E. W.: J. biol. Chem. **212**, 485 (1955).
5. GIUSTINA, G., TEMELCOU, O.: G. Biochim. **4**, 181 (1955).
6. ANDO, T., ISHII, S., HASHIMOTO, C., YAMASAKI, M., IWAI, K.: Bull. Chem. Soc. Japan **25**, 132 (1952).
7. ANDO, T., IWAI, K., YAMASAKI, M., HASHIMOTO, C., KIMURA, M., ISHII, S., TAMURA, T.: Bull. Chem. Soc. Japan **26**, 406 (1953).
8. ANDO, T., YAMASAKI, M., ABUKUMAGAWA, E., ISHII, S., NAGAI, Y.: J. Biochem. (Tokyo) **45**, 429 (1958).
9. HASHIMOTO, C.: Bull. Chem. Soc. Japan **28**, 385 (1955).
10. ANDO, T., NAGAI, Y., FUJIOKA, H.: J. Biochem. (Tokyo) **44**, 779 (1957).
11. ANDO, T., IWAI, K., KIMURA, M.: J. Biochem. (Tokyo) **45**, 27 (1958).
12. HASHIMOTO, C.: J. Chem. Soc. Japan, Pure Chem. Sect. (Nippon Kagaku Zasshi) **80**, 800 (1959).
13. ANDO, T., ABUKUMAGAWA, E., NAGAI, Y., YAMASAKI, M.: J. Biochem. (Tokyo) **44**, 191 (1957).
14. ANDO, T., YAMASAKI, M., ABUKUMAGAWA, E.: J. Biochem. (Tokyo) **47**, 82 (1960).
15. YAMASAKI, M.: Sci. Pap. Coll. Gen. Educ. Univ. Tokyo **9**, 31, 49 (1959).
16. FELIX, K., KREKELS, A.: Z. physiol. Chem. **295**, 107 (1953).
17. FELIX, K.: The chemical structure of proteins. Ciba Foundation Symp., 151, London: J. and A. Churchill 1953.
18. MONIER, R., JUTISZ, M.: Biochim. biophys. Acta (Amst.) **14**, 551 (1954).
19. FELIX, K., FISCHER, H., KREKELS, A., RAUEN, H. M.: Z. physiol. Chem. **286**, 67 (1950).
20. ŠORM, F., ŠORMOVA, Z.: Collect. Czech. Chem. Commun. **16**, 207 (1951).
21. FELIX, K.: Amer. Sci. **43**, 431 (1955).
22. FELIX, K. (translated into Japanese by ANDO, T. and IWAI, K.): Chemistry of proteins (AKABORI, S., MIZUSHIMA, S., Eds.), Vol. 5, p. 265. Tokyo: Kyoritsu Shuppan Co. 1957.
23. FELIX, K., HASHIMOTO, C.: Z. physiol. Chem. **330**, 205 (1963).
24. NUKUSHINA, J.: Dissertation for a degree of Rigaku-shi (Bachelor of Science) (Univ. Tokyo), March 1964, and NUKUSHINA, J., ISHII, S., ANDO, T.: to be published.

25 a. WALDSCHMIDT-LEITZ, E., KÜHN, K., ZINNERT, F.: Experientia (Basel) **7**, 183 (1951).

25 b. WALDSCHMIDT-LEITZ, E., PFLANZ, L.: Z. physiol. Chem. **292**, 150 (1953).

26. WALDSCHMIDT-LEITZ, E., VOH, R.: Z. physiol. Chem. **298**, 257 (1954).
27. SCANES, F. S., TOZER, B. T.: Biochem. J. **63**, 565 (1956).
28. ANDO, T., HASHIMOTO, C.: J. Biochem. (Tokyo) **45**, 453 (1958).
29. HASHIMOTO, C.: J. Chem. Soc. Japan, Pure Chem. Sect. (Nippon Kagaku Zasshi) **79**, 848 (1958).
30. ONOUE, K., OTA, S., MORISAWA, S., KAWACHI, T., FUJII, S.: Fukuoka Acta med. (Fukuoka Igaku Zasshi) **49**, 2709 (1958).
31. HIROHATA, R.: J. Japan. Biochem. Soc. (Seikagaku) **30**, 661 (1958).
32. OTA, S., MURAMATSU, M., HIROHATA, R., OKUDA, Y., YANG, C.-C., KAO, K.-C., CHIN, W.-C., CHANG, C.-C., IMAI, Y., ONO, T.: Ann. Rep. Lab. Protein Chem. Yamaguchi Medical School (in Japanese) **1**, 1 (1966).
33. WALDSCHMIDT-LEITZ, E., GUTERMANN, H.: Z. physiol. Chem. **344**, 50 (1966).
34. ANDO, T., TOBITA, T., YAMASAKI, M.: J. Biochem. (Tokyo) **45**, 285 (1958).
35. TOBITA, T., YAMASAKI, M., ANDO, T.: J. Biochem. (Tokyo) **63**, 119 (1968).
36. KAWANISHI, Y., IWAI, K., ANDO, T.: Proc. 9th Symp. on Protein Structure, 96, Osaka (Nov. 1958).
37. KAWANISHI, Y., IWAI, K., ANDO, T.: J. Biochem. (Tokyo) **56**, 314 (1964).
38. FELIX, K., LANG, A.: Z. physiol. Chem. **193**, 9 (1930).
39. FELIX, K.: Ber. ges. Physiol. **61**, 349 (1931).
40. DIRR, K., FELIX, K.: Z. physiol. Chem. **205**, 83 (1932).
41. FELIX, K., DIRR, K., HOFF, A.: Z. physiol. Chem. **212**, 50 (1932).

References to Table VI-1 (continued):

42. Waldschmidt-Leitz, E., Ziegler, Fr., Schäffner, A., Weil, L.: Z. physiol. Chem. **197**, 219 (1931).
43. Felix, K., Goppold-Krekels, A., Hübner, L., Yamada, T.: Acta biol. med. germ. **2**, 48 (1959).
44. Zimmermann, E.: Dissertation, Johann Wolfgang Goethe-Universität, Frankfurt a. M. 1959.
45. Tomášek, V.: Chem. Listy **50**, 840 (1956).
46. Nakano, M., Tobita, T., Ando, T.: Presented at 41th and 42th General Meetings of Japan. Biochem. Soc. (Tokyo, 28 Oct. 1968 and Hiroshima, 7 Oct. 1969); Abstr. in J. Japan. Biochem. Soc. (Seikagaku) **40**, 555 (1968), **41**, 441 (1969). Nakano, M., Tobita, T., Ando, T.: Biochim. biophys. Acta (Amst.) **207**, 553 (1970).

In fact, the peptide A·A and its anhydride were isolated as picrate salts[2] from the hydrolyzate (Gross, 1922; Kossel and Staudt, 1927). The peptide of M·M type, which is absent in the structural formula, was also supposed to be present in the hydrolyzate (Gross, 1922; Nelson-Gerhardt, 1919).

Although clupeine and salmine are known to be resistant to the action of pepsin, they are readily hydrolyzed by pancreatic enzymes. Waldschmidt-Leitz *et al.* [1929 (1), 1931, 1933, 1935 (1)] proposed the following structure for clupeine based on their finding of peptides of the types M·A, A·M, $(A \cdot M \cdot A)_2$, A·Pro·A, and A·A in the hydrolyzates of clupeine with various pancreatic enzymes. The structure is of the same type as proposed by Kossel]

H-MA-AMA-AMA-AProA-AM-AA-OH

Felix's group, on the other hand, reported the isolation of various peptides from the partial hydrolyzates of whole clupeine with trypsin-kinase (Felix *et al.*, 1932) and acid [Dirr and Felix, 1932 (2); Felix *et al.*, 1933]:

H-Arg·Arg-OH	H-(Val, Arg)-OH[3]
H-(Ser, Arg)-OH[3]	H-(Val?, Arg, Arg)-OH[3]
H-(Ala, Arg)-OH[3]	H-Arg·Arg·Arg·Arg-OH
H-Arg·Hyp-OH	

From these results Felix assumed a structure which comprised repeating sequences of MAA and AAM for clupeine (Felix and Mager, 1937), particularly taking into account the isolation of tetraarginine (Felix *et al.*, 1933):

H-ProAA-AAVal-SerAA-AAHyp-ValAA-AAPro-
AlaAA-AAPro-SerAA-AAVal-AlaAA-OH

After World War II, a large number of peptides were obtained and identified in partial hydrolyzates of several protamines by various research groups using the modern techniques developed in protein chemistry. These are summarized in Table VI-2. As can be seen from the upper part of the table, Felix considered the N-terminal sequence to be H-Pro·Ala·Arg-, which was slightly different from the previous structure. He finally proposed the following general formula for clupeine (from

[2] In those days arginine and its peptides were separated and identified as their picrate, picrolonate or flavianate salts.

[3] (X, Y) represents a peptide in which the sequence of the amino acids in parentheses (X and Y) is not determined, and so on.

Table VI-2. Peptides identified in partial hydrolyzates of unfractionated or nearly unfractionated protamines

Protamine	Hydrolysis method	Peptide isolated [References]
Clupeine-Me·HCl *(Clupea harengus)*	1 N HCl, boil for 8 h	Arg·Arg·Arg·Arg (recovery: 35% N/clupeine N); Arg·Arg·Arg (2.5%); Ser·Arg·Arg (6.7%); Arg·Arg + Arg (total 24.9%) [1]. Ala·Ala [2]. Ser·Ala; Ile·Ala; Val·Ala; Val·Ala·Arg·Arg; Pro·Val·Arg_3·Arg [3—5].
DNP-Clupeine *(Clupea harengus)*	HCl	DNP-Pro·Ala; DNP-Pro·Ala·Arg; DNP-Pro·Ala·Arg_4·(Thr, Ser)[a]; DNP-Pro·Ala·Arg_4·(Thr, Ser)·Arg_4·Val·Ile [3—5][a].
	HCl	DNP-Pro·Ala; DNP-Pro·Ala·Ser; DNP-Ser·Ala·Ser [6].
	Trypsin, pH 7.85, 37 °C, 18 h	DNP-Pro·Arg and DNP-Ala·Arg·Arg [7].
Clupeine *(Clupea harengus)*		Ala·Arg; Ala·Arg·Arg[a]; Ile·Arg; Ile·Arg·Arg[a]; Val·Arg; Ser·Arg; Pro·Arg; Ala·Ser·Arg; Pro·(Ala, Arg); Ala·Ser; Pro·Ser; Arg·Arg; Arg_3[a]; Arg_4[a] [8].
	Trypsin (M/16 HCl-treated), pH 8.0, 30 °C, 20.5 h	Thr·Thr·Arg (recovery: 0.32 mole/mole); Thr·Arg (0.28); Val·Ser·Arg (0.57); Pro·Val·Arg (0.11); Pro·Ile·Arg (0.04); Ala·Gly·Arg (0.34); Ser·Arg (0.17); Ala·Ser·Arg (0.09); Ala·Arg (1.18); Ser·Ser·Ser·Pro·Ile·Arg (0.26); Pro·Arg (0.29); Ala·Ser·Arg·Pro·Val·Arg (0.34); Arg (2.72); Arg·Pro·Arg (0.5); Ala·Arg·Arg (0.17); Arg·Arg (3.40) [7].
Clupeine *(Clupea pallasii)*	conc. HCl, 37 °C, 240 h	Arg (5.4_3 mole/mole)[b]; Arg·Val; Val·Arg; Thr·Arg (total of the above three, 1.3_9); Arg·Ala; Arg·Pro; Ser·Arg (total 1.8_2); Ala·Arg (0.24); Pro·Arg (0.25); Arg·Arg (3.8_6); Arg·Arg·Arg (0.28); Neutral fraction (total 3.0_8) [9].
	Trypsin, pH 7.81, 30 °C, 20 h	Arg (2.6 mole/mole)[b]; Thr·Arg (0.41); Ser·Arg (0.19); Ala·Arg (0.99); Pro·Arg (0.50); Arg·Arg (3.83); Thr·Thr·Arg (0.22); Val·Ser·Arg (0.70); Pro·Val·Arg (0.19); Pro·Ile·Arg (0.01_8); Ala·Gly·Arg (0.19); Ala·Ser·Arg (0.19); Arg·Pro·Arg (0.55); Arg·Ala·Arg (0.02_7); Arg·Thr·Arg (0.01_8); Ala·Arg·Arg (0.12); Pro·Arg·Arg (0.05_5); Arg·Arg·Arg (0.14); Arg·(Val, Ser)·Arg (0.02_7); Val·Ser·Arg·Arg (0.01_8); Arg·Thr·Thr·Arg (0.01_8); Arg·Arg·Pro·Arg (0.25); Ser·Ser·Arg·Pro·Ile·Arg (0.16); Ala·Ser·Arg·Pro·Val·Arg (0.46) [10, 11].
DNP-Clupeine *(Clupea pallasii)*	Trypsin, pH 7.6, 30 °C, 6—40 h	DNP-Ala·Arg; DNP-Pro·Arg; DNP-Ala·Arg·Arg; DNP-Pro·Arg·Arg [12—14].

Table VI-2 (continued)

Protamine	Hydrolysis method	Peptide isolated [References]
Salmine *(Oncorhynchus)* (and DNP-Salmine)	Trypsin; 2 N HCl, 100 °C, 3 h, or 1 N HCl, 37 °C, 8 days	Pro·Ile; Pro·Val; Val·Gly; Val·Ser; Gly·Gly; Arg·Val; Pro·Arg; Ile·Arg; Pro·Ile·Arg; Pro·Val·Arg; Gly·Gly·Arg; Val·Ser·Arg; Ala·Ser·Arg; Ser·Ser·Arg; Arg·Pro·Val; Pro·Val·Arg·Arg [15, 16][c].
DNP-Salmine *(Oncorhynchus)*	Formic acid-acetic anhydride-perchloric acid, or trypsin	DNP-Pro·Arg; DNP-Pro·Arg·Arg [15][c].
DNP-Salmine *(Oncorhynchus keta)*	Trypsin, pH 7.6, 30 °C, 6—40 h	DNP-Pro·Arg; DNP-Pro·Arg·Arg [12—14].
Salmine *(Oncorhynchus keta)*	Trypsin (diphenylcarbamyl chloride-treated), pH 8.0, 37 °C, 20.5 h	Val·Ser·Arg (0.83 mole/mole); Ile·Arg (0.18); Ser·Ser·Arg (0.24); Pro·Val·Arg (0.05_8); Ala·Ser·Arg (0.22); Pro·Arg (0.90); Gly·Gly·Arg (0.86); Ser·Ser·Ser·Arg·Pro·$\binom{\text{Val}}{\text{Ile}}$ Arg (0.21); Arg (1.78); Arg·(Pro, Val)·Arg (0.19_6); Arg·Pro·Arg and Pro·Arg·Arg (total 0.12); Arg·Ser·Ser·Arg (0.01_4); Arg·Arg (5.25); Arg·Gly·Gly·Arg (+); Arg·(Arg, Pro)·Arg (0.09_2); Arg·Arg·Arg (0.12) [17, 18].
Iridine *(Salmo irideus)*	Trypsin (diphenylcarbamyl chloride-treated), pH 8.0, 37 °C, 20.5 h	Val·Ser·Arg (0.76 mole/mole); Ser·Ser·Arg (0.21); Val·Arg (0.13); Ile·Arg (0.11); Ala·Ser·Arg (0.28); Gly·Gly·Arg (0.41); Ala·Arg (0.20); Ser·Ser·Ser·Arg·Pro·$\binom{\text{Val}}{\text{Ile}}$·Arg (0.17); Pro·Arg (0.48); Arg (0.93); Arg·(Pro, Val)·Arg (0.16); Arg·Val·Ser·Arg and Val·Ser·Arg·Arg (total 0.12); Arg·Pro·Arg (0.36); Arg·Ser·Ser·Arg (0.13); Arg·Val·Arg (0.05_6); Arg·Gly·Gly·Arg (0.19); Pro·Arg·Arg (0.50); Arg·Arg (4.54); Arg·(Arg, Pro)·Arg (0.24); Arg·Arg·Arg (0.44) [17,18].
Mugiline β[d] *(Mugil japonicus)*	2 N HCl, 100 °C, 3—40 h; 1 N NaOH, 100 °C, 20 h; 10.7 N HCl, 37 °C, 72—100 h; or trypsin, pH 7.8, 37 °C, 4 h	Ala·Pro; Pro·Ile; Ile·Pro; Val·Val[e]; Thr·Ser; Ala·Ala; Val·Thr; Pro·Val·Ile; Ala·Pro·Val·Ile; Ala·(Pro, Val); Ile·Arg; Ala·Arg; Arg·Ala; Ser·Arg; Arg·Glu; Val·Arg; Pro·Arg; Val·Val·Arg; Arg·(Thr, Ser); Ala·Pro·Val·Ile·Arg; Ala·(Pro, Ile)·Arg; Ala·(Pro, Ile)·Arg·Arg; Glu·(Ser, Pro, Ile)·Arg; Glu·(Ser, Pro, Ile)·Arg·Arg; Val·Arg_n?; Arg·Arg·Glu; Pro·Ile·Arg_{1-2}; Ser·Arg_n; Arg_2; Arg_3; Arg_4; Arg_5 [19—23][c].

[a] Identification is uncertain.

[b] M. W. of clupeine was assumed to be 4,360 as free base and 5,450 as sulfate.

[c] All the results obtained with each hydrolyzate are shown collected together.

[d] Unfractionated specimen or its nitro-derivative, or a main fraction βI obtained by some fractionation with countercurrent distribution procedure, was used for these experiments.

[e] An anhydride of this dipeptide was obtained in 1937 by Hirohata in a partial hydrolyzate of mugiline β by heating with 25% sulfuric acid in an oil bath for 16—20 h [Hirohata, R.: J. Biochem. (Tokyo) **25**, 519 (1937)].

References to Table VI-2:

1. FELIX, K., RAUEN, H. M., ZIMMER, G. H.: Z. physiol. Chem. **291**, 228 (1952).
2. FELIX, K.: Experientia (Basel) **8**, 312 (1952).
3. Identified by paper chromatography. FELIX, K.: The chemical structure of proteins. Ciba Foundation Symp., p. 151. London: J. and A. Churchill 1953.
4. Identified by paper chromatography. FELIX, K. (translated into Japanese by ANDO, T. and IWAI, K.): Chemistry of proteins (AKABORI, S., MIZUSHIMA, S., Eds.), Vol. 5, p. 265. Tokyo: Kyoritsu Shuppan Co. 1957.
5. FELIX, K., GOPPOLD-KREKELS, A., HÜBNER, L., MEISSNER, P.: Bull. Soc. Chim. biol. (Paris) **40**, 1973 (1958).
6. ŠORM, F., ŠORMOVA, Z.: Collect. Czech. Chem. Commun. **16**, 207 (1951).
7. Determined by ion-exchange column chromatography. NUKUSHINA, J.: Dissertation for a degree of Rigaku-shi (Bachelor of Science) (Univ. Tokyo), March 1964, and NUKUSHINA, J., ISHII, S., ANDO, T.: to be published.
8. Identified by paper chromatography. ISHIDA, S.: Fukuoka Acta med. (Fukuoka Igaku Zasshi) **46**, 389 (1955).
9. Determined by ion-exchange column chromatography. ANDO, T., ISHII, S., KIMURA, M.: Biochim. biophys. Acta (Amst.) **31**, 255 (1959).
10. Determined by ion-exchange column chromatography. ANDO, T., ISHII, S., YAMASAKI, M.: Biochim. biophys. Acta (Amst.) **34**, 600 (1959).
11. ISHII, S., YAMASAKI, M., ANDO, T.: J. Biochem. (Tokyo) **61**, 687 (1967).
12. ANDO, T., ABUKUMAGAWA, E., NAGAI, Y., YAMASAKI, M.: J. Biochem. (Tokyo) **44**, 191 (1957).
13. ANDO, T., YAMASAKI, M., ABUKUMAGAWA, E.: J. Biochem. (Tokyo) **47**, 82 (1960).
14. YAMASAKI, M.: Sci. Pap. Coll. Gen. Educ., Univ. Tokyo **9**, 49 (1959).
15. Identified by paper chromatography. MONIER, R., JUTISZ, M.: Biochim. biophys. Acta (Amst.) **14**, 551 (1954).
16. MONIER, R., JUTISZ, M.: Biochim. biophys. Acta (Amst.) **15**, 62 (1954).
17. WATANABE, S.: Thesis for Doctor of Science (Univ. Tokyo), June 1969.
18. ANDO, T., WATANABE, S.: Int. J. Protein Res. **1**, 221 (1969).
19. IKOMA, T.: J. Japan. Biochem. Soc. (Seikagaku) **26**, 13 (1954).
20. OTA, S., SHICHIRAKU, A.: Fukuoka Acta med. (Fukuoka Igaku Zasshi) **49**, 108 (1958).
21. ONOUE, K., OTA, S., MORISAWA, S., KAWACHI, T., FUJII, S.: Fukuoka Acta med. (Fukuoka Igaku Zasshi) **49**, 2709 (1958).
22. HIROHATA, R.: J. Japan. Biochem. Soc. (Seikagaku) **30**, 661 (1958).
23. OTA, S., ONOUE, K., HIROHATA, R., KAWACHI, T., OKUDA, Y., MORISAWA, S., FUJII, S.: Z. physiol. Chem. **317**, 1 (1959).

Clupea harengus, which is made up mainly of dipeptides of monoamino acids (M·M) alternating with tetrapeptides of arginine (FELIX, 1960)[4].

$$\text{H-Pro}\cdot\text{Ala}\cdot\text{Arg}_4\cdot\text{Thr}\cdot\text{Ser}\cdot\text{Arg}_4\cdot[\text{M}\cdot\text{M}\cdot\text{Arg}_4]_n\text{-OH}$$

On the basis of his experimental results, FELIX postulated that all components of clupeine (*i.e.*, several molecular species present as a mixture in a heterogeneous protamine like clupeine; see Chap. VII) have the same N-terminal sequence of 6—8 amino-acid residues and that the differences among the components must occur along the chains towards the C-terminals. He also stated that tri- or tetrapeptide sequences of monoamino acids may occur in a chain of components. Such a chain should contain a hexa- or octapeptide of arginine in order to maintain the ratio of arginine residues to monoamino acid residues at 2:1.

Some other kinds of N-terminal sequences, H-Pro·Ala·Ser- and H-Ser·Ala·Ser-, as shown in Table VI-2, have once been reported in a clupeine specimen by Czechoslovakian workers (ŠORM and ŠORMOVA, 1951).

---

[4] As described above (Chap. VI. B. 1), N-terminal proline and alanine were later reported to be present in this clupeine specimen (FELIX and HASHIMOTO, 1963).

French biochemists identified many peptides qualitatively on paper chromatograms of the partial hydrolyzates of salmine (described only as of the genus *Oncorhynchus*), as also shown in Table VI-2 [MONIER and JUTISZ, 1954 (1, 2)]. From their results, a general structure was suggested for salmine. This formula, in contrast to those described above, is composed of unique sequences of the form M·Arg$_n$, in addition to those of M$_2$·Arg$_n$.

$$\text{H-Pro}\cdot\text{Arg}_n\cdot\text{M}\cdot\text{Arg}_{n'}\cdot[\text{Arg}_{n''}\cdot\text{MM}]_6\cdot\text{Arg}_{n'''}\text{-OH}$$

A Japanese group obtained quantitative results for a number of tryptic peptides in clupeine from *Clupea pallasii*, as also shown in Table VI-2 (ANDO *et al.*, 1959; ISHII *et al.*, 1967). These results show that two thirds of all the monoamino acids in whole clupeine are present in the form of the sequence -Arg·M·M·Arg-, and one third in the sequences -Arg·M·Arg- and M·Arg- (the N-terminal sequence, hence M represents alanine or proline). Thus clupeine does not have such simple periodic structures as assumed by the German groups, the results being rather in agreement with the French workers' qualitative results for salmine. The complexity of the structure is further indicated by the presence of a unique sequence -Arg·M·M·Arg··M·M·Arg-. Such a sequence was found for the first time in protamines. It shows a single arginine residue separated only by neutral amino acids from other arginine residues. Furthermore, the presence of the peptides, Arg·Arg, Arg·Arg·Pro·Arg and Arg·Arg·Arg, in the hydrolyzate suggests the occurrence in clupeine of numbers of sequences containing three or more arginine residues.

Quite similar results have also been obtained recently with whole clupeine from *Clupea harengus* (NUKUSHINA, 1964; NUKUSHINA *et al.*, 1964), whole salmine from *Oncorhynchus keta*, and iridine from *Salmo irideus* (WATANABE, 1969; ANDO and WATANABE, 1969; WATANABE and ANDO, to be published). Thus, a few components having similar but not identical structures are expected to be present in each of these unfractionated protamines, clupeine, salmine, and iridine.

In their work with mugiline β from Formosan grey mullet (*Mugil Japonicus* TEMMINCK et SCHLEGEL), HIROHATA's group (HIROHATA, 1958; OTA *et al.*, 1959) found that unique sequences, unlike those of other protamines, occur in the chain. It appears to contain -M·M·M·M- and -M·M·M-, and also sequences of more than four arginine residues [IKOMA, 1954 (1)].

All or some of the components of these unfractionated protamines have recently been separated homogeneously and studied for their complete amino-acid sequences, in which all such sequences as found in Table VI-2 are indeed present (see Chap. VIII).

As regards the structural features of galline, a protamine obtained from fowl sperm, some recent results with both unfractionated galline and some fractionated components are described in Chap. VII. B. 6, and those of a basic protein from bull sperm heads in Chap. VII. B. 7.

Chapter VII

Heterogeneity of Protamines and Homogeneous Molecular Species of Protamines

A. Heterogeneity in Protamines

Histones and protamines are present in animal cell nuclei as basic nuclear proteins associated with deoxyribonucleic acid. They are known to be inhomogeneous and probably consist of definite amounts of several components which are very similar in their nature and structure. The significance of the heterogeneity of these basic nuclear proteins has not yet been completely elucidated, though some particular activity in controlling DNA-dependent biosynthesis is sometimes attributed to each component of the histones.

It was well known even before KOSSEL's time that clupeine is not homogeneous (GOTO, 1902; MIYAKE, 1927; KOSSEL and SCHENCK, 1928). The heterogeneity of protamine was revealed in earlier studies by tentative fractionation based on the differential solubility of presumptive components, although the fractionation was quite incomplete. In later studies protamines were shown to be inhomogeneous by physical methods, and in more recent years they have been fractionated into all or some of their homogeneous components by the use of various chromatographic procedures.

The picrate salt of clupeine or salmine was often separated into two or three fractions of slightly different amino-acid and nitrogen content by stepwise precipitation from the warm solution in 67% aqueous acetone at room temperature, 0° and −11 °C [RASMUSSEN, 1934; ANDO *et al.*, 1952, 1957 (1, 3)]. Sulfates of clupeine and salmine were separated from aqueous solution according to the difference in solubility into a few fractions, each having a different ratio of total N to α-amino N (WALDSCHMIDT-LEITZ *et al.*, 1931, 1951; WALDSCHMIDT-LEITZ and VOH, 1954). However, SCANES and TOZER (1956) reported that no fractionation was effected by a similar treatment. On the basis of the different solubility of the methyl ester hydrochloride of clupeine in methanol containing hydrochloric acid, several fractions were obtained stepwise which differed in amino acid and methoxyl content and in molecular weight (FELIX and DIRR, 1929; FELIX and MAGER, 1937; ŠORM and ŠORMOVA, 1951). Some of the fractions so obtained may possibly have been contaminated with degradation products of the protamine during the esterification and fractionation treatments.

Clupeine, or its ester, was reported to be inhomogeneous from the results of ultracentrifuge sedimentation analysis [DAIMLER, 1952 (1, 2); RAUEN *et al.*, 1952] and electrophoresis (RAUEN *et al.*, 1952; ZIMMERMANN, 1959). On the other hand, specimens of clupeine and salmine tentatively fractionated from their acid salts by means of solubility differences were sometimes reported to be homogeneous on the basis of sedimentation-velocity measurements (VELICK and UDENFRIEND, 1951) and electrophoresis (VELICK and UDENFRIEND, 1951; UI and WATANABE, 1953; UI, 1956). According to UI, the clupeine specimen gave a simple symmetrical pattern on electrophoresis while salmine in a phosphate buffer appeared to be inhomogeneous, reversibly forming aggregates. He supposed that an electrophoretic procedure could not distinguish components of a protamine which might be fairly similar to each other.

The countercurrent distribution technique showed protamines, including clupeine, salmine, iridine and mugiline β, to be inhomogeneous [RAUEN *et al.*, 1953; FELIX *et al.*, 1957; RASMUSSEN, 1963; ANDO *et al.*, 1957 (1, 3); SCANES and TOZER, 1956; ANDO and SAWADA, 1961; MORISAWA, 1957; IKOMA, 1954 (2)]. In this case, protamine esters, *i.e.*, the methyl ester hydrochloride of clupeine (RAUEN *et al.*, 1953) and of mugiline β (MORISAWA, 1957), were separated into three and two fractions, respectively, while unesterified protamine sulfates were hardly fractionated by the technique, though patterns of a heterogeneous nature were obtained.

Before 1960, all structural studies of protamines had been performed with wholly unfractionated or only partially fractionated specimens. The C-terminal residues of most of the protamines studied were occupied exclusively by arginine and the N-terminal residues mainly by proline, while the evidence for the presence of both proline and alanine in whole clupeine obtained from Pacific herring [ANDO *et al.*, 1955, 1957 (1, 2), 1958 (2)], indicated the presence of more than two molecular species in clupeine.

During the following decade, actual isolation of one or more homogeneous components of clupeine and some other protamines was achieved by means of chromatographic procedures. The English workers, SCANES and TOZER (1956) introduced a method combining countercurrent distribution and chromatographic techniques for the fractionation of protamines. They fractionated a commercial specimen

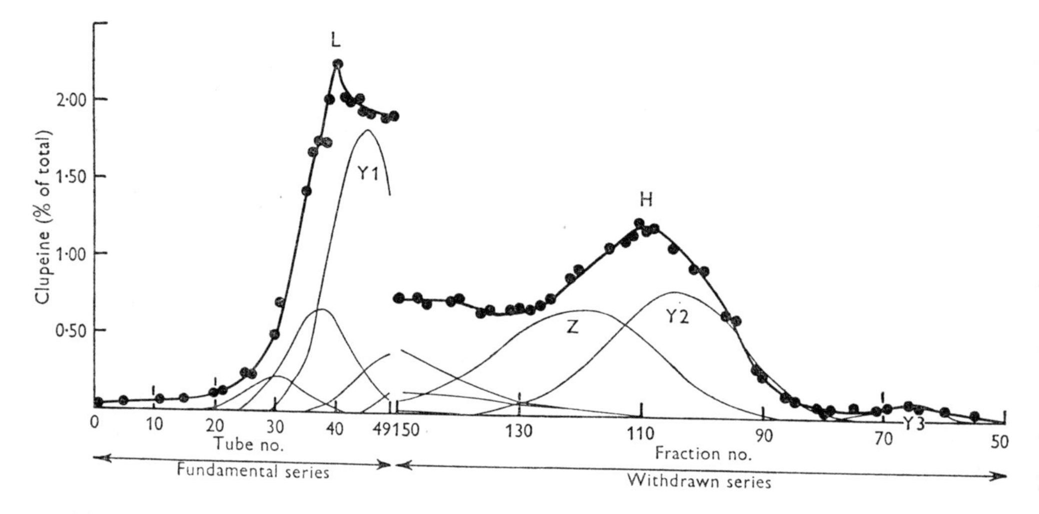

Fig. VII-1. A 150-transfer distribution (50-tube machine) of crude clupeine in *n*-propanol-3 M sodium acetate (1.5:1, v/v). ● —— ●, experimental; ——, calculated (from SCANES, F. S., and B. T. TOZER, 1956)

of whole clupeine of unknown species by means of preparative countercurrent distribution (solvent system: *n*-propanol *vs.* 3 M sodium acetate: 1.5:1.0, v/v) into two fractions, L (main, Y1) and H (main, Y2) (Fig. VII—1). At the same time they obtained separately a third fraction, Z, by elution chromatography of whole clupeine on a column of active alumina, not pretreated with buffer, using 0.5 M phosphate (K_2HPO_4) buffer as eluant (Fig. VII—2). Similar results were reported with a commercial specimen of salmine.

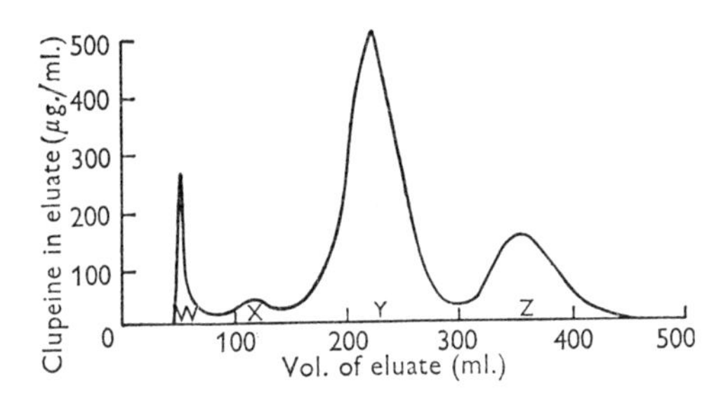

Fig. VII-2. Elution chromatography of crude clupeine sulphate (50 mg) from a column of 40 g of alumina (type O) in a tube of 1.2 cm internal diam. Developing solvent: 0.5 M-K_2HPO_4 (from SCANES, F. S., and TOZER, B. T., 1956)

Japanese biochemists fractionated clupeine prepared from *Clupea pallasii* into two fractions, Y and Z, by elution chromatography on a column of alumina which had been buffered with the eluting solvent (0.45 M or 0.48 M K_2HPO_4, pH 9.0—9.2) (ANDO and SAWADA, 1959, 1961). On subsequent application of countercurrent distribution (solvent system: *n*-butanol *vs.* 0.12 M sodium *p*-toluenesulfonate containing 0.40 M NaCl), the Y fraction was further separated into YI and YII, while the Z fraction was found to be homogeneous (see Chap. VII. B).

The characteristics of these three tentative fractions of clupeine ,as obtained by the English and Japanese workers, are set out in Table VII—1, together with three fractions obtained by a German group (RAUEN *et al.*, 1953; FELIX *et al.*, 1957) by means of countercurrent distribution (solvent system: 5% lauric acid in *n*-butanol *vs.* 15% aqueous sodium acetate), and one fraction obtained by another German group (WALDSCHMIDT-LEITZ and PFLANZ, 1953; WALDSCHMIDT-LEITZ and VOH, 1954) by means of fractionation based on solubility in water. The last column of the table gives for comparison one probably homogeneous fraction, corresponding to Z, obtained by RASMUSSEN (1963).

In spite of some differences in the species of fish used and the methods employed, the fractions designated i, ii, and iii in Table VII-1 correspond to one another as regards their amino-acid content. However, there are differences in the N-terminal residues. Fraction iii, which has the simplest amino acid composition with N-terminal alanine, was the first to be purified and have its complete amino-acid sequence elucidated (ANDO *et al.*, 1962). New methods for the fractionation of clupeine have recently been developed and are described below (this chapter, Section B).

The problem remains to be solved, whether such heterogeneity in protamines can be ascribed to different stages of maturation of fish testes, differences between individuals of the same species, or different preparation procedures.

In an attempt to trace the origin of the heterogeneity, clupeine and iridine specimens were prepared under exactly the same mild conditions from completely mature testes of individual specimens of *Clupea pallasii* (ANDO and SAWADA, 1959, 1962) and *Salmo irideus* (ANDO and SAWADA, 1959, 1960), respectively. They used three Pacific herring caught at three different places off Hokkaido and three rainbow trout obtained

Table VII-1. Comparison of clupeine fractions obtained by different authors in their early studies

Fraction of corresponding composition	(A) SCANES and TOZER [1][a]			(B) ANDO and SAWADA [2][b]				(C) FELIX *et al.* [3][c]			(D) WALDSCHMIDT-LEITZ *et al.* [4][c]	(E) RASMUSSEN [5][c]
	(i)	(ii)	(iii)	(i)	(ii)	(iii)		(i)	(ii)	(iii)	(iii)	(iii)
Fraction	L (Y1)	H (Y2)	Z	YI	YII	Z	Z [2 a]	II	III	I	One fraction	One fraction
Amino-acid composition: Arg	81.5	85.5	83.1	(+ + +)	(+ + +)	(+ + +)	21	25	25	25	(+ + +)	20.3
Pro	6.2	8.3	4.7	(+)	(+)	(+)	2	2	4	2	(+)	2.1
Ala	3.6	3.7	4.2	+ +	+ +	+ +	3	2	2	3	(+)	2.9
Ser	6.3	4.7	5.9	+ +	+	+	3	3	3	2	(+)	2.1
Val	0.4	4.3	3.4	±	+ +	+ +	2	—	2	2	(+)	1.9
Thr	3.9	2.8	0	+	+	—	—	2	1	—	—	—
Ile	1.2[d]	0.2[d]	0[d]	+	±	—	—	1	—	—	—	—
Gly	1.4	0.1	0	+	±	—	—	1	—	—	—	—
N-Terminal amino acid	Pro (main) and Ala (minor)		Ala (main)	Ala(76%) and Pro(24%)	Pro(85%) and Ala(15%)	Ala(100%)		Pro	Ser(main) Thr(minor) Ala(minor)	Pro	Pro	

[a] Commercial specimen of clupeine from herring of unknown species. Amino-acid contents (amino acid g/100 g protein) were estimated by column chromatography on starch (MOORE and STEIN, 1948, 1949).

[b] Clupeine from Pacific herring *(Clupea pallasii)*. Amino-acid contents were estimated by densitometry of multiply developed paper chromatograms. (+ + +) indicates the presence in a large amount and (+) in an appreciable amount, though not estimated. + + and + in appreciable amounts when estimated, ± only a trace and — absence.

[c] Clupeine from Norwegian Sea herring *(Clupea harengus)*.

[d] Ile and/or Leu.

1. SCANES and TOZER, 1956.
2. ANDO and SAWADA, 1961.

2 a. ANDO *et al.*, 1962; AZEGAMI *et al.*, 1970. Amino-acid contents (moles/mole) were estimated by an automatic analyzer, and from the amounts of amino acids present in the tryptic peptides as well as from the molar ratio of the amino terminus: proline: arginine.

3. FELIX *et al.*, 1957; moles/mole.
4. WALDSCHMIDT-LEITZ and PFLANZ, 1953; WALDSCHMIDT-LEITZ and VOH, 1954.
5. RASMUSSEN, 1963; moles/mole.

from a hatchery near Tokyo. The chromatographic behavior, N-terminal amino acids and amino acid composition of the specimens were compared within each group.

All the specimens of clupeine and iridine proved to be chromatographically heterogeneous, as can be seen from the two examples of clupeine presented in Fig. VII-3, which gave patterns not different from that of clupeine from pooled fish material. By N-terminal analysis of each clupeine specimen, proline and alanine were demonstrated to be present always in approximately the same molar ratio,

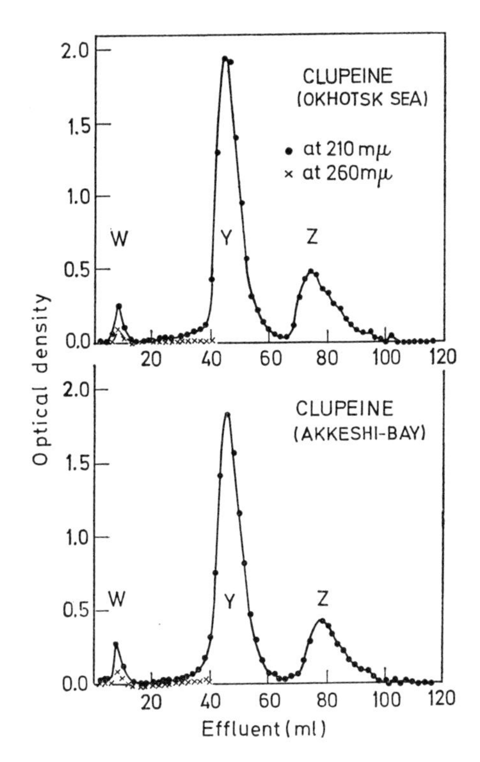

Fig. VII-3. Alumina chromatography of crude clupeine hydrochloride prepared from two individuals, obtained at Okhotsk Sea (upper) and in Akkeshi Bay (lower) around Hokkaido (from ANDO, T., and F. SAWADA, 1962)

0.42:0.58, evidently indicating the heterogeneity of the N-terminal. No difference was found in the amino-acid composition of the various clupeine specimens. All the specimens of iridine had exclusively proline at the N-terminus.

Thus the heterogeneity of clupeine and iridine obtained from individual fish was confirmed, and it was concluded that heterogeneity is an intrinsic property of protamines. Its possible biological significance in the basic nuclear proteins, protamine and histone, seems to be of interest and importance but remains to be clarified (see Chap. X. C. 1).

As to the heterogeneity of other protamines, mugiline β from grey mullet *(Mugil japonicus)* was studied by chromatography on a column of alumina, calcium phosphate gel, or CM-cellulose (OTA, 1961; OTA *et al.*, 1966). Hardly any homogeneous component was obtained from whole mugiline β by these procedures except that mugiline β was fractionated by column chromatography on buffered alumina according to the method of ANDO and

SAWADA (1960) to yield a probably homogeneous fraction giving a single peak (OTA *et al.*, 1966). The fraction had the following molar ratios of amino acid residues: Thr_1:Ser_1:Glu_1:Pro_3:Ala_2:Val_2:Ile_2:Arg_{38}, with N-terminal proline only, and a calculated molecular weight of 7,165 as free base. The heterogeneity of mugiline β specimens prepared separately from a large number of individual fish was also recognized by chromatographic patterns, while a probably homogeneous pattern was obtained only for the protamine specimen prepared from a fish just before spawning (OTA *et al.*, 1966).

Thynnine from tunny fish was shown to be heterogeneous by partition chromatography on a column of Sephadex G-50 using a solvent system of *n*-propanol/3 M sodium acetate. The fractionation of the protamine by column chromatography on buffered alumina was not complete (BRETZEL, 1967, 1971). Recently it was separated into four fractions, thynnine Y1, Y2, Z1 and Z2 by column chromatography on CM-Sephadex (BRETZEL, 1971) (see Chap. VII. B. 5).

A protamine, galline, obtained from sperm cell nuclei of domestic fowl, has recently been separated into several fractions by chromatography on a column of Bio-Gel CM-30 (NAKANO *et al.*, 1968, 1969 and 1970) (see this Chap. VII. B. 6).

B. Preparative Fractionation of Protamines into Their Components

1. Fractionation of Whole Clupeine into Y and Z Fractions by Column Chromatography on Buffered Alumina (Homogeneous Clupeine Z)

Elution chromatography on alumina, introduced by SCANES and TOZER (1956) for the fractionation of clupeine, was slightly modified to give better resolution by eluting the protein from a buffered alumina column (ANDO and SAWADA, 1960).

The specimen of clupeine sulfate (200.9 mg) was adsorbed on a column (1.88 × 46 cm) of alumina buffered with 0.48 M phosphate buffer[1], pH 9.0—9.2, and eluted with the same buffer at 25 °C at the flow rate of 14 ml/20 min/tube. Two main peaks, Y and Z, were obtained. The latter was tentatively divided into two parts, ZI and ZII as shown in Fig. VII-4. Each of these three fractions gave a single peak at the expected position after rechromatography on a buffered alumina column as shown in Fig. VII-5 a, b, c (ANDO and SAWADA, 1961).

Although the fraction Y showed a single peak on rechromatography, it proved to be far from homogeneous on the N-terminal analysis. The fraction Y was further separated into two fractions, YI and YII, by countercurrent distribution with a solvent system consisting of *n*-butanol and aqueous sodium *p*-toluenesulfonate (0.12 M) containing NaCl (0.40 M) (Fig. VII-6 a). Fractions YI and YII were still not completely homogeneous since the analysis of their N-terminals gave 24% proline and 76% alanine for YI and 85% proline and 15% alanine for YII). In contrast, no further fractionation was achieved with the fraction Z by countercurrent distribution (Fig. VII-6 b shows ZII). Both ZI and ZII (see Fig. VII-4) gave 100% alanine on their N-terminal analysis and had the same amino-acid composition: Ser, Pro, Ala, Val and Arg (ANDO and SAWADA, 1961). The specimen of clupeine Z was again obtained later by elution chromatography followed by rechromatography of the Z fraction (ANDO *et al.*, 1962; AZEGAMI *et al.*, 1970).

Thus the Z fraction was thought to be homogeneous and used for the structural analyses, described later, which further confirmed the homogeneity.

[1] Alumina for chromatography (Wako Pure Chemicals Co. Ltd.), sieved between 200 to 300 mesh, was washed 5 times with tapwater and 5 times with distilled water (adjusting pH of the supernatant to 4.3—4.7 with HCl), dried at 110 °C, and calcinated for 3—4 h over a Teclu burner. The alumina was then rinsed twice with 0.48 M K_2HPO_4, pH 9.0—9.2, and left overnight. The precipitate suspended in the buffer was transferred into a column.

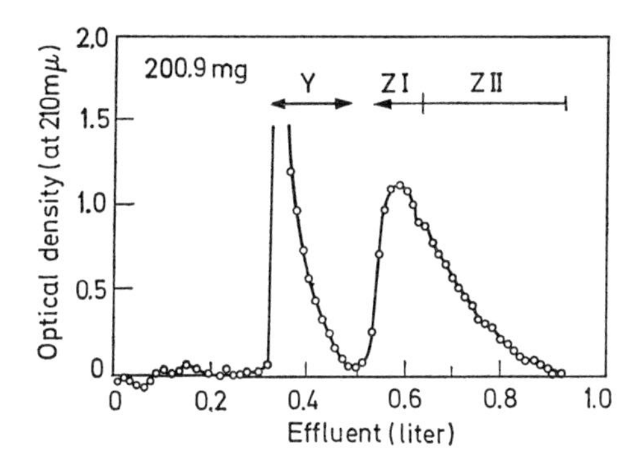

Fig. VII-4. Preparative chromatography of clupeine sulfate using a buffered column. Alumina-0.48 M K_2HPO_4, pH 9.0—9.2 at 25 °C (from ANDO, T., and F. SAWADA, 1961)

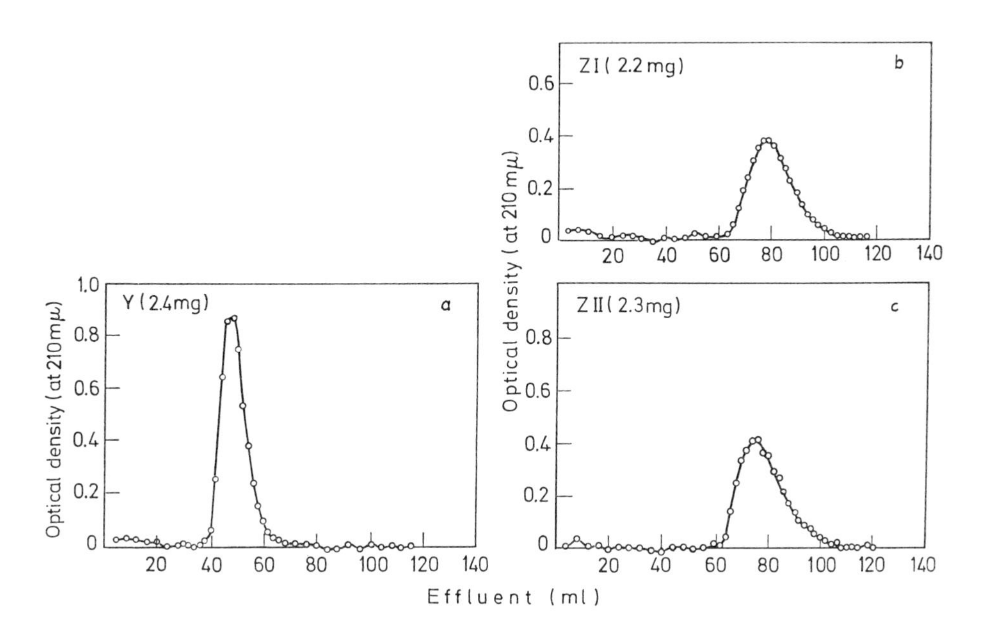

Fig. VII-5. Analytical rechromatography of clupeine fractions Y, ZI and ZII using buffered columns, resulting in chromatograms a, b and c, respectively. Sulfate specimens (Y, ZI and ZII fractions) obtained by preparative chromatography using buffered columns (see Fig. VII-4) were eluted from buffered alumina with 0.48 M K_2HPO_4, pH 9.0—9.2 at 25 °C (from ANDO, T., and SAWADA, F., 1961)

Desalting and Recovery of Clupeine Y and Z Hydrochlorides from the Chromatographic Fractions

Each of the combined fractions Y and Z obtained from clupeine sulfate (513 mg) was neutralized to pH 6—7 by adding 1 N HCl. Each solution was adsorbed on a small column (1.5 × 3—6 cm) of Amberlite CG-50 (Type II) which had been buffered at pH 7 with 0.48 M

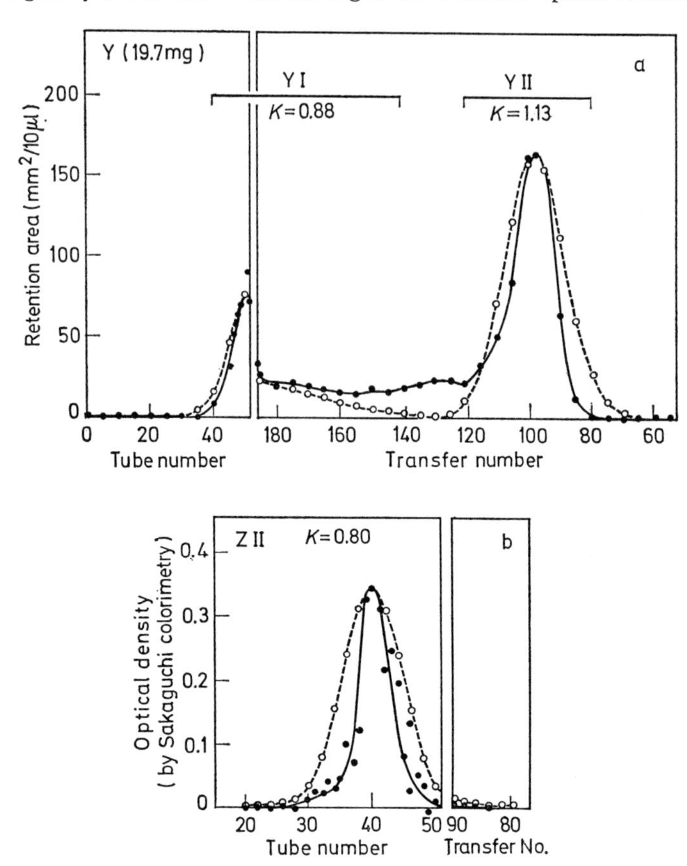

Fig. VII-6. Countercurrent distribution of clupeine fractions Y (a) and ZII (b) obtained by preparative chromatography using a buffered column (Fig. VII-4). Solvent system: 0.12 M aqueous sodium *p*-toluenesulfonate-0.40 M NaCl *vs.* *n*-butanol, 22—23 °C. —— ● —— experimental curve; --- ○ --- theoretical curve. *K*: Distribution coefficient (from ANDO, T., and SAWADA, F., 1961)

K_2HPO_4. The column was washed with ice-cooled 0.5 M acetic acid for desalting according to the method of ISHII [1960 (2)]. Then after washing with water, the protamine adsorbed on the resin was eluted with cold 0.1 N HCl. The acid effluent was neutralized with Amberlite JRA-400 (HCO_3^-) and the solution was concentrated, added to cold 0.1 N HCl until pH 5—6, and lyophilized. Thus clupeine Y hydrochloride (ca. 300 mg) which was still a mixture, and a homogeneous component, clupeine Z hydrochloride (ca. 100 mg) were obtained with overall recoveries of more than 90% of the arginine residues by SAKAGUCHI colorimetry (AZEGAMI, 1961; AZEGAMI *et al.*, 1970).

2. Fractionation of Whole Clupeine into Y and Z Fractions by Column Chromatography on CM-Cellulose (Homogeneous Clupeine Z)

Since the reproducibility of chromatography using alumina is not always satisfactory, chromatographic systems using various ion exchangers were examined for the fractionation of whole clupeine.

Among these a chromatographic system using CM-cellulose (0.60 or 0.63 meq/g, Serva Entwicklungslabor, Heidelberg) was found reproducible and satisfactory for the fractionation of whole clupeine into Y and Z fractions in place of the method of adsorption column chromatography on alumina [Suzuki and Ando, 1968 (2)]. The chromatographic pattern obtained is quite similar to that given by alumina. This system of chromatography, with its high reproducibility and recovery as well as its faster flow rate, will also be useful for the fractionation of other protamines or histones.

Fraction Z (the second peak) contained only a homogeneous component, clupeine Z, as shown by analysis of the amino-acid composition and the N-terminus (100% alanine), while fraction Y (the first peak) was a mixture. On further fractionation of fraction Y on a CM-cellulose column using several buffer systems over a wide pH range, two fractions were obtained: YI-rich (80—90% N-terminal alanine contaminated with 10—20% N-terminal proline) and YII-rich (80—90% N-terminal proline mixed with 10—20% of N-terminal alanine). Complete separation of the components YI and YII was unsuccessful by this procedure.

3. Fractionation of Trinitrophenylated (TNP-) Whole Clupeine into TNP-YI, TNP-Z and Free YII Components by Column Chromatography on CM-Cellulose (Homogeneous TNP-Clupeine YI, TNP-Clupeine Z and Free Clupeine YII)

Clear separation of the clupeine Y fraction into two components, YI and YII, is now desired. 2,4,6-Trinitrobenzene sulfonic acid (TNBS) is known to selectively modify amino groups in a peptide chain to yield trinitrophenyl (TNP) derivatives but never to react with imino groups (Okuyama and Satake, 1960). Thus, the N-terminal alanine residue of clupeine YI and Z should be trinitrophenylated while the N-terminal proline of clupeine YII is not, so that modification of whole clupeine with TNBS will give TNP-clupeine YI, TNP-clupeine Z and free clupeine YII. Some differences are then expected in the power of a resin to adsorb free and trinitrophenylated protamine components.

TNP-YI, TNP-Z and free YII were indeed successfully separated by one-step elution of modified whole clupeine from a CM-cellulose column as shown in Fig. VII-7 [Ando and Suzuki, 1966; Suzuki and Ando, 1968 (2)]. The amino-acid composition of TNP-clupeine Z agrees with that estimated for homogeneous clupeine Z (Ando *et al.*, 1962; Azegami *et al.*, 1970). TNP-alanine is known to yield free alanine during acid hydrolysis (Okuyama and Satake, 1960). The molar ratios of the amino-acid content of free clupeine YII and TNP-clupeine YI proved to be simple integral numbers in the structural studies (see Chap. VIII. B and C), in which accurate analysis of the amino-acid composition and the N-terminus confirmed their homogeneity.

4. Fractionation of Whole Clupeine into Its Three Components, YI, YII and Z, and of Whole Salmine and Iridine into Some of Their Components by One-Step Elution Chromatography on a Column of CM-Sephadex or Bio-Gel CM (Homogeneous Clupeine YI, YII and Z; Homogeneous Salmine AI, Iridine I and II)

Unmodified whole clupeine from Pacific herring has now been fractionated successfully on a preparative scale by one-step elution from a column of CM-Sephadex C-25 or Bio-Gel CM-2 into its three components, clupeine YI, YII and Z as shown in Fig. VII-8 (A) (Ando and Watanabe, 1969). The homogeneity of each component was confirmed by analysis of the amino-acid composition and the N-terminus. This method of fractionation has proved to be more clear-cut, reproducible and quantitative than any of the methods previously employed for clupeine. Though the reason for such fractionation by the chromatographic system is not clear, it could be due to differences in the nature of the N-terminal residues at pH 5.8 during

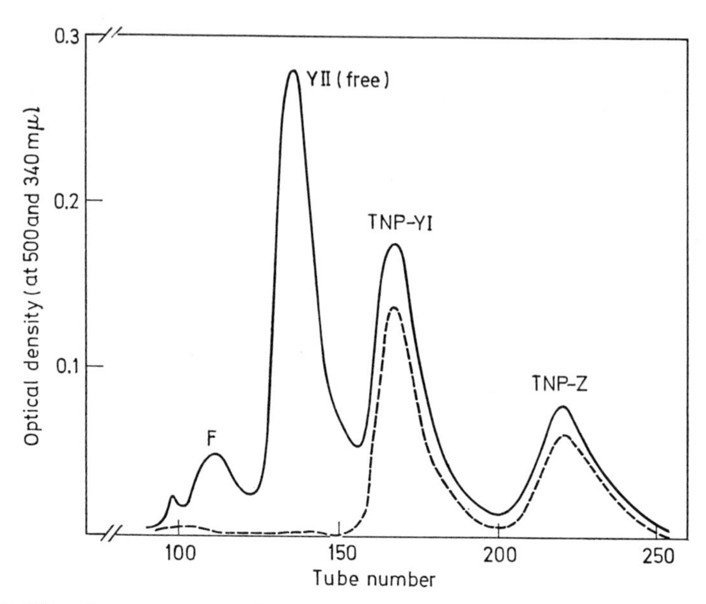

Fig. VII-7. The chromatographic fractionation of TNP-ated clupeine into TNP-clupeine YI, TNP-clupeine Z and free clupeine YII. TNP-ated whole clupeine (8.1 mg) was added on a column of CM-cellulose (0.9 × 89 cm). It was developed with 0.1 M acetate buffer, pH 3.51, containing 0.6 M sodium chloride. The flow rate was 2.3 ml per fraction per 10 min at room temperature. The column was covered with aluminium foil to protect the TNP-derivatives from photodecomposition. The effluent from the column was analyzed by SAKAGUCHI colorimetry ($OD_{500\,m\mu}$) and spectrophotometry at 340 mμ. —— $OD_{500\,m\mu}$ SAKAGUCHI color value. --- $OD_{340\,m\mu}$ Absorption of TNP-group [Reproduced from SUZUKI, K., and ANDO, T., 1968 (2)]

elution, as well as in the characteristics along the peptide chains among the three components.

Four different batches of clupeine specimens from *Clupea pallasii* were separated by means of this chromatographic system. The fractions of the peaks of YI, YII and Z obtained were 34 ± 6%, 36 ± 5% and 26 ± 4%, respectively. The amount of F, a minor fraction that was eluted before the YII fraction (cf. Figs. VII-7, 8), was 4 ± 3%. Thus about 96% of whole clupeine is composed of the three main components, YI, YII and Z, and their average ratio in whole clupeine is $1:1:0.7_5$ [SUZUKI and ANDO, 1968 (2)].

Whole clupeine from *Clupea harengus* was also fractionated by chromatography into three components, clupeine Y'I, Y'II and Z' (see Chap. VIII. E) (CHANG, NAKAHARA and ANDO, to be published), just like clupeine from *Clupea pallasii*.

In the case of salmine and iridine, the chromatographic system can so far separate only some homogeneous components (ANDO and WATANABE, 1969), leaving the remaining components as a mixture. The total number of components of these protamines is unknown.

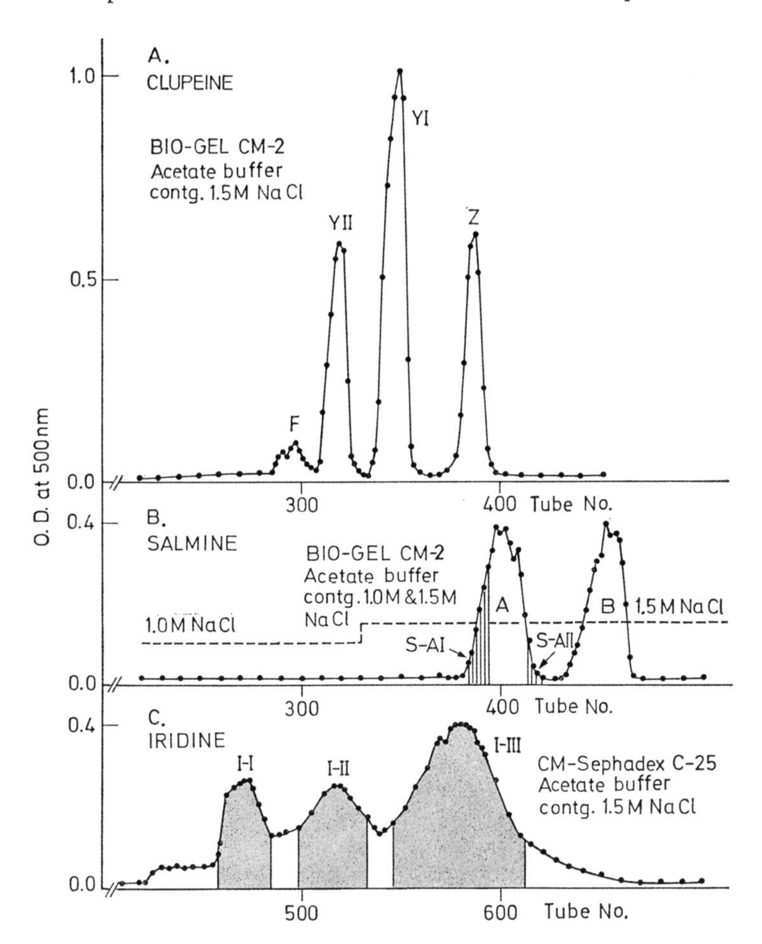

Fig. VII-8. Column chromatographic fractionation on preparative scale of protamines into all or some of their components using CM-Sephadex C-25 or Bio-Gel CM-2. A. Protamine: ca. 25 mg of whole clupeine sulfate (from *Clupea pallasii*) in a small amount of water. Column: 0.9 × 130 cm of Bio-Gel CM-2 (high capacity). Elution: 0.05 M acetate buffer, pH 5.8, containing 1.5 M NaCl at room temperature at a flow rate of 2.4 ml/h. B. Protamine: ca. 50 mg of whole salmine sulfate (from *Oncorhynchus keta*) in a small amount of water. Column: 0.9 × 150 cm of Bio-Gel CM-2 (high cap.). Elution: 0.05 M acetate buffer, pH 5.8, containing 1.0 M and 1.5 M NaCl at room temperature at a flow rate of 10 ml/h. C. Protamine: ca. 600 mg of whole iridine sulfate (from *Salmo irideus*) in as small as possible amount of the eluting buffer containing 0.5 M NaCl. Column: 3.0 × 146 cm of CM-Sephadex C-25. Elution: 0.05 M acetate buffer, pH 5.8, containing 1.5 M NaCl at room temperature at a flow rate of 85 ml/h (Reproduced from ANDO, T., and WATANABE, S., 1969)

Salmine from *Oncorhynchus keta* was separated into two main fractions, A and B, by chromatography on a column of Bio-Gel CM-2 or CM-Sephadex C-25 (Fig. VII-8, B). Fraction A, though eluted in a single peak, consisted of two main components, S-AI and S-AII, which occupied mainly the first half and the second half of this peak, respectively. Rechromatographed S-AI (composition: Arg_{21}, Ser_4, Pro_3, Gly_2, Val_2) and S-AII (composition: Arg_{21} or Arg_{22}, Ser_4, Pro_3, Gly_2, Ala_1, Ile_1) were present in fraction A in a molar ratio of approximately 2 to 1.

Iridine from *Salmo irideus* was separated by column chromatography on CM-Sephadex C-25 into three main fractions (Fig. VII-8, C). Each of these was rechromatographed to yield I-I (composition: Arg_{22}, Ser_4, Pro_3, Gly_2, $Val_{1.3}$, $Ile_{0.7}$,)[2], I-II (Arg_{21}, Ser_4, Pro_2, Gly_2, Ala_1, Val_2) and I-III (though this fraction was still a mixture, it had an amino-acid composition similar to that of salmine B fraction).

One component of salmine (S-AI) and two components of iridine (I-I[2] and I-II) were used for structural studies (see Chap. VIII. F).

5. Fractionation of Whole Thynnine into Four Fractions, Y1, Y2, Z1 and Z2, by Chromatography on a Column of CM-Sephadex (Homogeneous Thynnine Y1, Y2, Z1 and Z2)

Whole thynnine sulfate prepared from the isolated sperm cell nuclei of tunny *(Thynnus thynnus)* (WALDSCHMIDT-LEITZ and GUTERMANN, 1966) was fractionated into four fractions, W, X, Y, and Z (BRETZEL, 1967), by chromatography on buffered alumina according to the method used for clupeine and iridine. These four fractions were not homogeneous, although the Z fraction was at first thought to be so.

Recently, whole thynnine sulfate was separated into four fractions, Y1, Y2, Z1 and Z2, as shown in Fig. VII-9 (BRETZEL, 1971), by column chromatography on

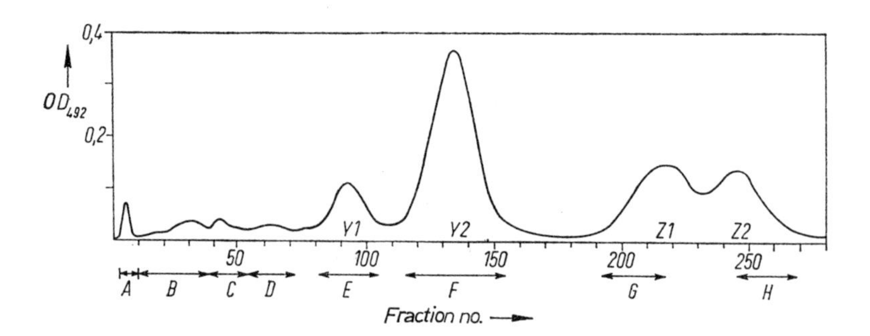

Fig. VII-9. Fractionation of whole thynnine by column chromatography on CM-Sephadex C-25. Column: 1.7 × 123 cm of CM-Sephadex C-25 (cap. 0.76 meq/g, Deutsche Pharmacia, Frankfurt/Main). Protamine: 75 mg of whole thynnine sulfate. Elution: 0.05 M sodium acetate buffer, pH 5.8, containing 1.3 M NaCl and 0.02% sodium azide at a flow rate of 18 ml/2 fr./h (from BRETZEL, G., 1971)

CM-Sephadex C-25 as used earlier for clupeine, salmine, and iridine (ANDO and WATANABE, 1969). The homogeneity of each fraction, after rechromatography, was confirmed by analysis of the amino-acid composition and the N-terminus, and by disc electrophoresis.

Thynnine Y1 and Y2 have the same amino-acid composition, $Arg_{20\pm1}$, Thr, Ser_2, Glx, Pro_2, Ala_3, Val_3, Tyr. Thynnine Z1 has the composition $Arg_{21\pm1}$, Thr, Ser_3, Pro_2, Ala, Val_4, Tyr, while Z2 is $Arg_{21\pm1}$, Thr, Ser_3, Pro_2, Ala_2, Val_3, Tyr. It is

[2] This fraction proved later to have two components, I-I a and I-I b, in a molar ratio of 1 to 2, as shown by structural studies on peptides in thermolysin and Neutral Protease digests of I-I.

suggested on the basis of an unpublished result from work with thermolytic peptides that the difference between the Y1 and Y2 fractions is that the latter has Gln and the former has Glu for Glx, respectively. Thus the Y1 fraction seems to be merely a product of the decomposition of the Y2 component during preparation. All these four fractions have the same terminal amino acids, proline at the N-terminus and a block of four arginine residues at the C-terminal region. The molecular weight of whole thynnine was previously estimated by the ultracentrifugal method to be 6,800 as sulfate (WALDSCHMIDT-LEITZ and GUTERMANN, 1966). The molecular weights (as Arg_{20}) of the fractionated components Y2, Z1 and Z2 are calculated from the amino-acid composition to be approximately 4,770, 4,486 and 4,458 as free bases, or ca. 5,750, 5,516 and 5,488 as sulfates, respectively.

The complete amino-acid sequences of the components Y1 and Y2, together with a preliminary sequence of the component Z1, have been determined recently, as shown below [BRETZEL, private communication, Dec. 1971; BRETZEL, 1972 (1, 2, 3)].

Y2: $\overset{1}{\text{H-Pro}}\cdot(\text{Arg})_4\cdot\overset{6}{\text{Gln}}\cdot\text{Ala}\cdot\text{Ser}\cdot\text{Arg}\cdot\text{Pro}\cdot\overset{10}{\text{Val}}\cdot(\text{Arg})_5\cdot\text{Tyr}\cdot(\text{Arg})_2$-
$\overset{20}{\text{Ser}}\cdot\text{Thr}\cdot\text{Ala}\cdot\text{Ala}(\text{Arg})_5\cdot\overset{30}{\text{Val}}\cdot\text{Val}\cdot(\text{Arg})_4\overset{34}{\text{-OH}}$

Y1: $\overset{1}{\text{H-Pro}}\cdot(\text{Arg})_4\cdot\overset{6}{\text{Glu}}\cdot\text{Ala}\cdot\text{Ser}\cdot\text{Arg}\cdot\text{Pro}\cdot\overset{10}{\text{Val}}\cdot(\text{Arg})_5\cdot\text{Tyr}\cdot(\text{Arg})_2$-
$\overset{20}{\text{Ser}}\cdot\text{Thr}\cdot\text{Ala}\cdot\text{Ala}(\text{Arg})_5\cdot\overset{30}{\text{Val}}\cdot\text{Val}\cdot(\text{Arg})_4\overset{34}{\text{-OH}}$

Z1: $\overset{1}{\text{H-Pro}}\cdot(\text{Arg})_?\cdot\text{Ser}\cdot\text{Ser}\cdot\text{Arg}\cdot\text{Pro}\cdot\text{Val}\cdot(\text{Arg})_?\cdot\text{Tyr}\cdot(\text{Arg})_2\cdot\text{Ser}$-
$\text{Thr}.\text{Ala}.\text{Val}.(\text{Arg})_?.\text{Val}.\text{Val}.(\text{Arg})_?\text{-OH}$

6. Fractionation of Whole Galline into Several Components by Column Chromatography on Bio-Gel CM-30

A protamine, galline, has been obtained from sperm cell nuclei of fowl *(Gallus domesticus)* and partially characterized (DALY *et al.*, 1951; FISCHER and KREUZER, 1953). There is a considerable difference in the amino-acid composition of the galline specimens obtained by DALY *et al.* and by FISCHER and KREUZER. DALY *et al.* report that whole galline contains arginine and histidine as basic amino acids and hence it must be a kind of diprotamine, whereas the German group finds that it is a typical monoprotamine since it contains only arginine as a basic amino acid. No further fractionation has been carried out by either group.

Recently a Japanese group (NAKANO *et al.*, 1968, 1969, 1970) prepared galline as the hydrochloride from sperm nuclei of New Hampshire fowl and studied some of its characteristics.

The amino-acid composition of whole galline obtained from 7- and 10-month old cocks is shown in Table VII-2 together with that obtained by the other two groups mentioned above. Serine, alanine, and arginine were found as the N-terminal residues of whole galline, and arginine and tyrosine as the C-terminal residues. An approximate molecular weight of whole galline was tentatively estimated to be 7,000 by the result of gel filtration on Sephadex G-25, G-50 and G-75. Whole galline hydrochloride gave four main bands by analysis on disc electrophoresis using acrylamide gel in 8 M urea at pH 4.5.

They attempted further fractionation of whole galline on a preparative scale by column chromatography on Bio-Gel CM-30 (low capacity) and obtained about 8 fractions (G-I—G-VIII) as shown in Fig. VII-10.

Table VII-2. Amino-acid composition of whole galline
The results obtained by three research groups (American, German and Japanese) are compared

	Galline			
	mole %		molar ratio[a]	
	NAKANO *et al.*, 1968, 1969, 1970 7-month old	10-month old	DALY *et al.* (1951)	FISCHER, KREUZER (1953)
Lys	0.7	0.9	—	—
His	0.3	0.3	2	—
Arg	66.0	55.4	45	42
Asp	0.4	0.5	1	—
Thr	1.6	1.6	3	2
Ser	10.1	16.9	14	5
Glu	0.5	0.7	2	1
Pro	2.4	3.5	7	5
Gly	7.5	8.7	10	1
Ala	3.2	3.5	4	5
Val	1.9	2.2	2	3
Met	0.1	—	—	—
Ile	—	—	1	1
Leu	0.2	0.2	—	—
Tyr	5.0	5.8	6	—
Total	(100)	(100)	97	65

[a] Ile was taken as standard for calculation.

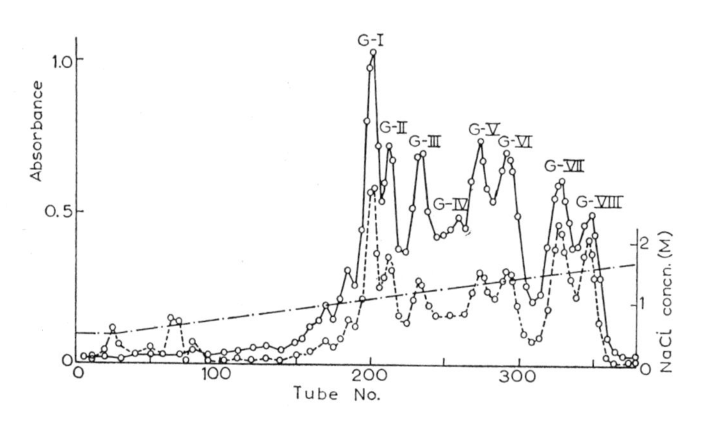

Fig. VII-10. Chromatography of whole galline hydrochloride on Bio-Gel CM-30. Sample: whole galline hydrochloride 92.3 mg. Column: 1.8 × 134 cm. Eluent: 0.1 M acetate buffer, pH 5.9, containing linear-gradient concentration of NaCl. Flow rate: 4.2 ml/tube/12 min 30 sec (from NAKANO, M., TOBITA, T., and ANDO, T., 1970)

The behaviour of each fraction on disc electrophoresis in 8 M urea at pH 4.5 was as follows. Fractions G-I—G-III gave 2 bands each, G-IV—G-VI 1 band each, G-VII 1 main and 2 minor bands, and G-VIII 1 main and 3 minor bands. The amino-acid composition, and

Table VII-3. Amino-acid composition and N- and C-terminal amino acids of galline fractions G-V and G-VI obtained nearly homogeneously from whole galline by chromatography on a Bio-Gel CM-30 column (NAKANO *et al.*, 1970)

	G-V		G-VI	
	mole %	Approx. residue per mole	mole %	Approx. residue per mole
Arg	56.1	24	62.8	29
Thr	2.7	1	1.6	1
Ser	21.2	9	16.6	8
Pro	5.6	2	4.3	2
Gly	5.4	2	5.3	2
Ala	4.7	2[a]	4.3	2[a]
Val	—	—	0.3	—
Tyr	4.3	2	4.8	2
Total	(100.0)	42	(100.0)	46
MW (calc.)		6,224		7,004
MW (calc. as HCl salt)		7,100		8,063
N-Terminal amino-acid[b]	Only Ala (recovery: 0.60 mole/mole)		Only Ala (recovery: 0.63 mole/mole)	
C-Terminal amino-acid[c]	Only Arg (recovery: 0.40 mole/mole)		Only Arg (recovery: 0.59 mole/mole)	

[a] Taken as standard for calculation.
[b] Determined by the DNP and DNS methods.
[c] Determined by hydrazinolysis.

N- and C-terminal amino acids were studied. From the results of these chemical and electrophoretic analyses, it was considered that some fractions were nearly homogeneous and others still mixtures.

Two fractions, G-V and G-VI, which appeared to be homogeneous, were studied more closely. They were rechromatographed and the amino-acid composition, and N- and C-terminal amino acids were determined as shown in Table VII—3. Some amino-acid sequences in the N- and C-terminal regions have also been deduced from the results of digestion with leucine aminopeptidase and carboxypeptidase B followed by A, respectively. Further, tryptic, chymotryptic, and thermolytic peptides of fraction G-V have been investigated.

Thus the total amino-acid sequence of the fraction G-V, together with a preliminary structure of a purified fraction G-I and a partial structure of G-VI, have been determined recently, as shown below (NAKANO *et al.*, 1972) (A = Arg).

G-V: H-Ala·A·Tyr·A·Ser(5)·Gly·A·Ser·A·Ser(10)·A·A·Thr·A·A(15)·A·A·Ser·Pro·A(20)·Ser·A·Gly·A·Ser(25)·Pro·A·A·A·A(30)·Ser·A·A·A·A(35)·A·Tyr·Gly·Ser·Ala(40)·A·A(42)-OH

G-I: H-Ser·A·Ser·Gly·Gly(5)·Val·A·A·A·A(10)·Tyr·Gly·Ser·A·A(15)·A·A·A·A·A(20)·A·A(22)·Tyr-OH

G-VI: H-Ala·A_n·Tyr·A·A·(A_{2-n}, Ser, Gly)-
{A·Thr·A, (A·Ser·A)$_5$, (A·Ser·Pro·A)$_2$, A·Gly·A, $A_{3\pm1}$}-
A·Tyr·Gly·(Ala, Ser)·A·A·A-OH

A precise comparison of the structural features of all the protamine components from fowl with those from fish sperm nuclei will be of great interest and importance from various points of view.

7. Isolation and Partial Characterization of a Basic Protein from Bull Sperm Heads

It has been proved impossible to extract basic protein from mammalian spermatozoa by the usual methods. This has led to the "sponge hypothesis" advanced by BRIL-PETERSEN and WESTENBRINK (1963), according to which the whole nucleus of the mammalian spermatozoon is regarded as a sponge consisting of a keratinoid network, held together by numerous disulfide bonds and associated with DNA to make the protein resistant against extraction or destruction.

Recently the isolation of the nuclear protein has been successfully performed after disruption of all the disulfide bridges in the postulated keratinoid network (COELINGH *et al.*, 1969; COELINGH, 1971).

The isolated heads of bull spermatozoa were reduced in a strongly denaturing medium (5 M guanidinium chloride, 0.28 M 2-mercaptoethanol, 0.005 M EDTA, and 0.5 M Tris · HCl, pH 8.5), leading to complete dissolution of the nucleus. Reoxidation of the released sulfhydryl groups was prevented by blocking them with ethylenimine. DNA was then removed from the cooled solution by acid precipitation (5M HCl and 2.5 M guanidinium chloride). After dialysis, fractionation of the dissolved head proteins by gel filtration on Sephadex G-75 yielded two fractions, as shown in Fig. VII-11.

The minor fraction (25%, peak I in Fig. VII-11) was of cytoplasmic origin and had no basic character. The major fraction (75%, peak II) was of nuclear origin and was rich in arginine and half-cystine. After further purification of fraction II by chromatography on Amberlite CG-50 (2 × 10 cm, 0.2 M acetic acid, then 0.05 M HCl),

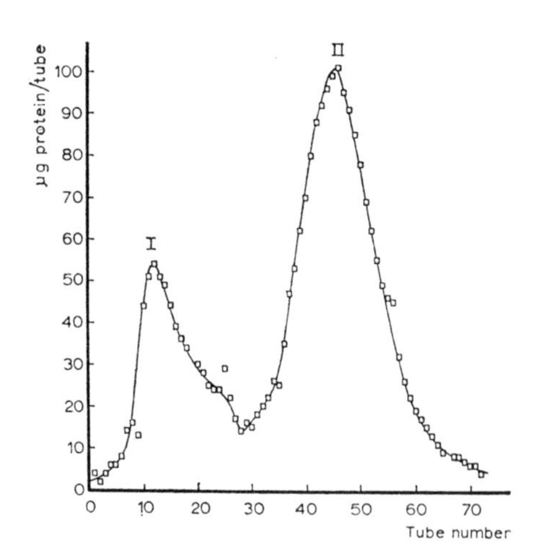

Fig. VII-11. Chromatography of proteins from bull sperm heads on Sephadex G-75. Sample: 2.6 mg of head protein. Column: 1.75 × 40 cm. Eluent: 0.01 M HCl-0.025 M NaCl. Flow rate: 13 ml/h. Fraction: 0.6 ml/tube (from COELINGH, J. P., ROZIJN, T. H., MONFOORT, C. H., 1969)

the basic protein was highly pure, as demonstrated by amino-acid analysis, acrylamide-gel electrophoresis and sedimentation behaviour, as well as by analyses of the N-(only Ala) and C-(only Gln) terminal amino acids. The following composition was deduced: Ala-(Arg_{24}, Cys_6, Thr_3, Ser_2, Gly_2, Val_2, Tyr_2, Leu_1, Ile_1, Phe_1, His_1)-Gln.

The molecular weight, about 6,200, is calculated from the formula, and is in accord with a value of about 7,000 estimated by the Archibald method. From the results of studies of the thermolytic peptides, it was suggested that the protein has a structural feature in which the central part of the molecule is very rich in arginine and at least half of the half cystine residues are located in the more terminal parts.

The nuclear basic protein of bull spermatozoa is thus considered to be a new class of histone (COELINGH, 1971), rich in arginine (resembling protamines in this respect) and cystine. The nucleus of the bull spermatozoon, therefore, may consist of a network of identical histone molecules linked by disulfide bridges and associated with DNA.

The complete amino-acid sequence of the basic protein has been deduced recently from the results of EDMAN degradation and of investigation of peptides obtained from the reduced and S-modified protein by degradation with thermolysin, chymotrypsin, and N-bromosuccinimide (COELINGH *et al.*, 1972). It is shown below:

H-$\overset{1}{\text{Ala}}$·Arg·Tyr·Arg·Cys·Cys·Leu·Thr·His·Ser·Gly·Ser·Arg·Cys·$(Arg)_7$·Cys-$(Arg)_6$·Phe·Gly·$(Arg)_6$·Val·Cys·Tyr·Thr·Val·Ile·Arg·Cys·Thr·Arg·$\overset{47}{\text{Gln}}$-OH.

Chapter VIII

Chemical Structure of Homogeneous Molecular Species of Protamines

A. Determination of the Complete Amino-Acid Sequence of Clupeine Z

The complete amino-acid sequence of clupeine Z was determined by the Tokyo group (ANDO *et al.*, 1962; ANDO, 1964; AZEGAMI *et al.*, 1970; IWAI *et al.*, 1971).

1. Amino-Acid Composition, Amino Terminus, and Molecular Weight

A rechromatographed specimen of clupeine Z obtained by column chromatography on alumina was subjected to structural analyses. Table VIII-1 shows the

Table VIII-1. Amino-acid composition of clupeine Z (Amino-acid N as a percentage of total N)

	Ser	Pro	Ala	Val	Arg	Recovery (%)
(1)[a]	3.35	2.21	3.51	2.20	88.7	—
(2)[b]	3.06	2.30	3.14	2.34	89.6	100.4
Mean value	3.2_1	2.2_6	3.3_3	3.2_7	89.2	
Molar ratio[c]	2.8_9	2.0_4	3.0_0	2.0_4	20.1 ± 0.6[d]	

[a] The acid hydrolyzate (6 N HCl, 100 °C, 48 h) of clupeine Z was analyzed in a Beckman/Spinco automatic amino-acid analyzer. Since these values were calculated from the data in mole percentage, the value for recovery is not shown.

[b] These data were obtained by summing the amounts of each amino acid present in tryptic peptides (see Table VIII-2 below). The nitrogen content of clupeine Z·HCl was 22.6 ± 0.2% by the Kjeldahl method.

[c] Alanine is taken as 3.00.

[d] A final value of 21 was taken for the total number of arginine residues in clupeine Z based on a separate experiment on the molar ratio of the amino terminus : proline : arginine (see text).

amino-acid composition of clupeine Z, as determined by the use of an automatic amino-acid analyzer, and as calculated from the results of analysis of the tryptic peptides.

Only DNP-Ala was found after acid hydrolysis of DNP-clupeine Z and the absence of any DNP-Pro terminus was shown by the calculated R value ($OD_{390}/OD_{360} = 0.53$) in 1 N HCl of DNP-clupeine Z [ANDO *et al.*, 1958 (2)]. The molecular weight of clupeine Z was estimated from the amount of DNP group introduced to be 4,400 as a free base or 5,500 as its hydrochloride.

The number of arginine residues was estimated more precisely by independent analyses in which the molar ratio, amino terminus (DNP method): Pro (CHINARD, 1952, for the hydrolyzate): Arg (SAKAGUCHI, 1951, for the hydrolyzate) was determined as 1.00:2.00:20.9.

Thus, the molecular formula $Ala_3\ Ser_3\ Pro_2\ Val_2\ Arg_{21}$ was obtained for clupeine Z. The molecular weight calculated from the formula is 4,163 as a free base or 4,929 as its hydrochloride.

2. Analysis of Tryptic Peptides of Clupeine Z, Leading to a Partial Chemical Structure

Employing a procedure essentially similar to that used for the structural study of unfractionated clupeine (ANDO *et al.*, 1959; ISHII *et al.*, 1967), 16 peptides and arginine were identified and determined quantitatively in a tryptic digest of clupeine Z as shown in Fig. VIII-1. Analyses of these tryptic peptides are shown in Table VIII-2. The kinds and amounts of partial amino-acid sequences present in clupeine Z are summarized in Table VIII-3.

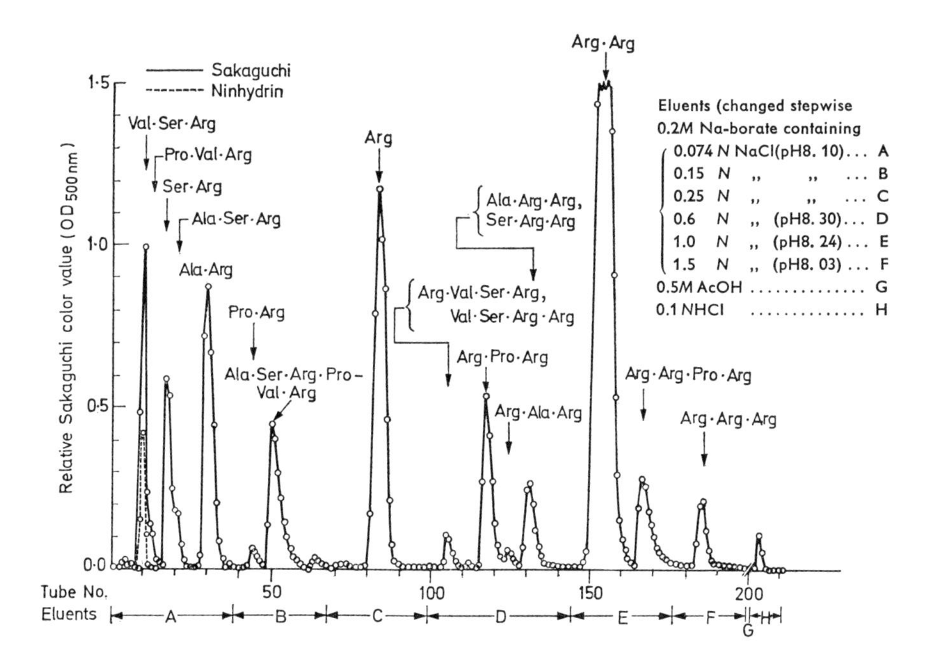

Fig. VIII-1. Column chromatographic separation of tryptic peptides of clupeine Z. The 20-h tryptic hydrolyzate of clupeine Z (approximately 10 µmoles) was chr ,natographed on Amberlite CG-50 (1.0 × 30 cm column). Elution was performed with an increasing NaCl concentration in 0.2 M sodium borate buffer stepwise as indicated in the figure at 30 °C and at a flow rate of 2.99 ml/tube/45 min. The color yield was 98 % (from ANDO *et al.*, 1962 and AZEGAMI *et al.*, 1970)

Table VIII-2. Analysis of peptides obtained from a tryptic hydrolyzate of clupeine Z (10.1 μmoles, 893 μmoles N)

Peaks obtained by column chromatographic separation (Fig. VIII-1) of tryptic peptides in a 20-h digest of clupeine Z were analyzed for their amino-acid composition, amino termini, etc. to determine the structure and amounts of peptides involved

Peak	Amino-acid composition (μmoles)							Amino	Arg/	R[a]	Structure	Amount[b]
No.	Arg	Ser	Val	Pro	Ala	Gly	Ile	terminus	Amino terminus			(μmoles)
3 (+ 3')	11.6	8.5	11.6	2.1				Val Pro	1.0	0.61	Val·Ser·Arg Pro·Val·Arg	9.3 2.3
4	7.1	6.8	0.0_4		0.1	0.05		Ser	1.1	0.54	Ser·Arg	7.1
4'	1.9	2.1			1.4		0.0_7	Ser Ala	1.1	0.56	Ser·Arg Ala·Ser·Arg	0.5 1.4
5	15.2				15.0			Ala	1.0	0.57	Ala·Arg	15.2
6	0.8	0.0_6	0.0_4	0.4	0.1			Pro ?	1.1	0.78	Pro·Arg	0.4
7	15.3	8.2	8.0	7.9	7.9			Ala	2.1	0.56	Ala·Ser·Arg·Pro·Val·Arg	7.7
7'	0.6	0.3	0.3	0.4	0.3			Ala	1.5	0.55	Ala·(Ser, Arg, Pro, Val)·Arg	0.3
8	25.0							Arg	0.9	0.53	Arg	25.0
9	2.0	0.6_5	0.8_4					Arg Val	2.0	0.59	Arg·(Val, Ser)·Arg Val·(Ser, Arg)·Arg	0.5 0.5
10	12.6		0.0_8	6.1	0.4_3			Arg	1.8	0.55	Arg·Pro·Arg	6.3
11	1.2				0.3_7			Arg	1.8	0.55	Arg·Ala·Arg	0.6

Table VIII-2 (continued)

Peak No.	Amino-acid composition (μmoles)							Amino terminus	Arg/ Amino terminus	R[a]	Structure	Amount[b] (μmoles)
	Arg	Ser	Val	Pro	Ala	Gly	Ile					
12	7.0	0.7			2.4			Ala Ser	2.2	0.60	Ala·Arg·Arg Ser·Arg·Arg	2.7 0.8
13	82.6							Arg	1.8	0.54	Arg·Arg	41.3
14	9.6			3.6				Arg	2.6	0.56	Arg·(Arg, Pro)·Arg	3.2
15	5.6							Arg	3.3	0.53	Arg·Arg·Arg	1.9
16	1.7	+		+	+	+			24	0.54		
Total amino acid (μmoles)	199.8	27.3	20.9	20.5	28.0							

[a] The R value represents the ratio, OD_{390}/OD_{360} of DNP-derivative in 1 N HCl.

[b] The molar amount of each peptide was calculated from the arginine content of the peak except for Pro·Arg, the amount of which was calculated from the proline content. The amount of mixed peptides in the peaks 3(+ 3'), 4' and 12 was calculated from the amino-acid composition. In the peak 9, it was estimated from the arginine content and the ratio of the DNP-N-terminals. Small amounts of aminv-acids other than the main one were neglected.

Table VIII-3. Partial amino-acid sequences found in clupeine Z (10.1 μmoles)

Amino-acid sequence	Amount (μmoles)
H·Ala·Arg-, -Arg·Ala·Arg-	18.5
-Arg·Val·Ser·Arg-	10.3
-Arg·Ala·Ser·Arg·Pro·Val·Arg-	9.5—10.3
-Arg·Ser·Arg-	8.4
-Arg·Pro·Arg-	9.9
Arg_n ($n \geq 2$)	+[a]

[a] The presence of a variety of arginine sequences is, of course, to be expected, but their individual characteristics cannot be determined from the results of tryptic digestion, since several peptide bonds between arginine residues have been cleaved by the enzyme.

As can be seen from Table VIII-3, satisfactory amounts (approximately 85 to 100%) of peptides were recovered; these included the sequences of all the monoamino-acids present among the arginine residues in clupeine Z. This indicates that each such sequence occurs once in a molecule. The following partial structure was given for clupeine Z:

H-Ala·Arg·[Arg·Ser·Arg, Arg·Ala·Ser·Arg·Pro·Val·Arg,
Arg·Pro·Arg, Arg·Val·Ser·Arg, Arg·Ala·Arg, 8 Arg]·Arg-OH

The location of arginine or arginine blocks in the clupeine molecule will, however, have to be determined in some other way.

3. Application of N → O Acyl Rearrangement Reaction to Clupeine Z, Followed by Selective Chemical Cleavage of the Chain, Leading to the Final Primary Structure

Tryptic hydrolysis can hardly be expected to give accurate information on the number of consecutive arginine residues in a protamine molecule. Thus, a complementary method of partial hydrolysis had to be developed so as to obtain longer peptide fragments in which intact sequences of arginine are retained.

The N → O acyl rearrangement reaction at the hydroxyamino-acid residues (Ser and Thr) and the selective chemical cleavage of the chain at the resulting ester bonds have been studied in detail using unfractionated clupeine (Iwai, 1959, 1960, 1961; Iwai and Ando, 1967). This method of degradation retains intact sequences of arginine residues in the products.

Since a clupeine Z molecule contains three residues of serine but no threonine, four peptides should be obtained by the rearrangement reaction (as illustrated below in Chart VIII-3). Thus the N → O acyl shift induced at the serine residues by treating the protamine with concentrated sulfuric acid at 20 °C for 4 days gave rearranged clupeine as an amorphous powder on addition of absolute ether to the solution at —10 °C. The product was then subjected to selective chemical fission of the chain at the resulting ester bonds by acid

(6 N HCl at 20 °C for 16 h) or weak alkali (0.3 N Na_2CO_3 at 30 °C for 4 h) after protection of exposed amino groups (those of the three serine residues and the original one of the N-terminal alanine) by acetylation, because, when the amino groups are kept free, the reverse reaction is accelerated by alkali.

Fractionation of the acid degradation products was not satisfactory (NAKAHARA *et al.*, 1967), and further studies are required.

The alkaline degradation products were satisfactorily fractionated by stepwise elution chromatography on an Amberlite CG-50 column to yield four main peaks consisting as expected of acetylpeptides (Fig. VIII-2). The structure of each frag-

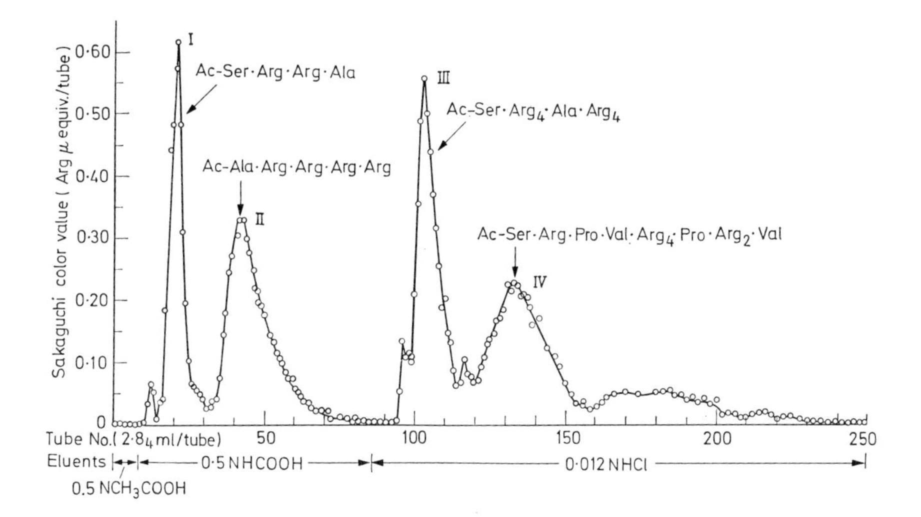

Fig. VIII-2. Chromatographic fractionation and identification of acetyl oligopeptides obtained by selective chemical degradation products from clupeine Z (2.2 μmoles). Amberlite CG-50 column (0.95 × 30 cm), flow rate of 5.5 ml/h, 30 °C. Recovered color value, 90% (tube No. 1—250) (from ANDO *et al.*, 1962; IWAI *et al.*, 1971)

ment was worked out first, as indicated in Table VIII-4, by determination of the amino-acid composition and the C-terminal, and finally, as demonstrated in Fig. VIII-2, by stepwise degradation with carboxypeptidases A and B. Digestion of fragments III and IV by the enzymes is shown in Chart VIII-1 for examples in which liberated amino acids were determined by an amino-acid analyzer, while the amino-acid composition and the C-terminus of acylated core peptides were determined by the analyzer and hydrazinolysis, respectively.

The sequence of fragments I—IV in a clupeine Z chain can be inferred as indicated in the last column of Table VIII-4, by considering the N-(Ala) and C-(Arg) terminals of the protamine together with the amino-acid sequences of tryptic peptides, especially the hexapeptide (H-Ala·Ser·Arg·Pro·Val·Arg-OH). Thus the complete amino-acid sequence of clupeine Z was worked out as shown in Chart VIII-2.

The structure given for clupeine Z was further supported quite satisfactorily and quantitatively by the results obtained by the action of leucine aminopeptidase (Fig.

Table VIII-4. Analysis of oligopeptides obtained by the selective fission of rearranged clupeine Z

Chrom. fraction	Amino-acid composition (mol. ratio)	C-Terminus	Structure	Peptide sequence
I	Arg 2.2_2 Ser 1.0_5 Ala 1.0_0	Ala	Ser·$(Arg)_2$·Ala	2nd
II	Arg 4.0_6 Ala 1.0_0	Arg	Ala·$(Arg)_4$	N-terminal
III	Arg 8.1_0 Ser 0.9_7 Ala 1.0_0 (Val 0.0_6)	Arg	Ser·$(Arg)_n$·Ala·$(Arg)_{8-n}$	C-terminal
IV	Arg 6.9_3 Ser 1.0_7 Pro 2.0_0 Val 2.0_9	Val	Ser·Arg·Pro·Val·$(Arg)_m$·Pro·$(Arg)_{6-m}$·Val	3rd

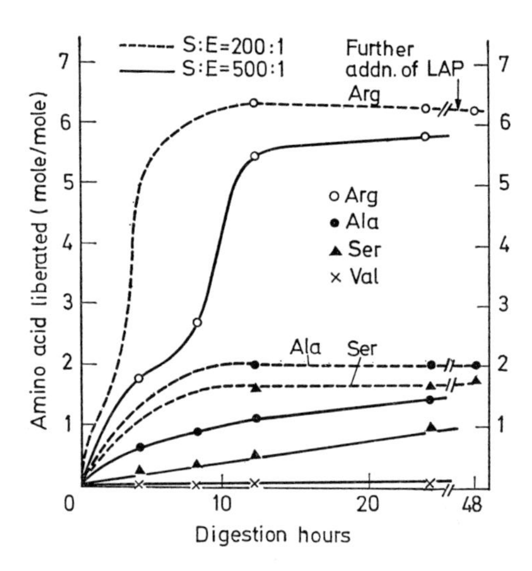

Fig. VIII-3. Hydrolysis of clupeine Z by leucine aminopeptidase (from Ando, 1964; Azegami *et al.*, 1970)

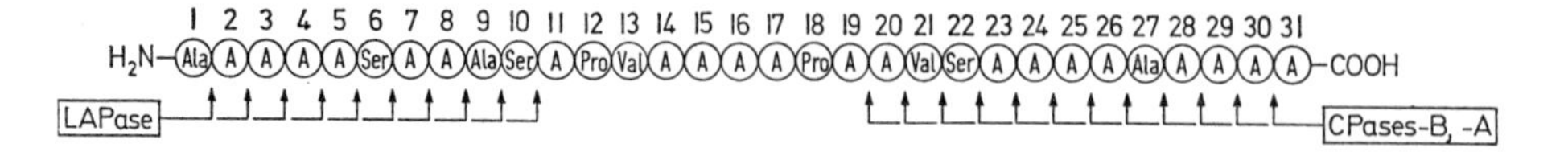

Chart VIII-2. The complete amino-acid sequence of clupeine Z (A = Arg). Modes of digestion by leucine aminopeptidase (LAPase) and carboxypeptidases (CPases) B and A are also illustrated

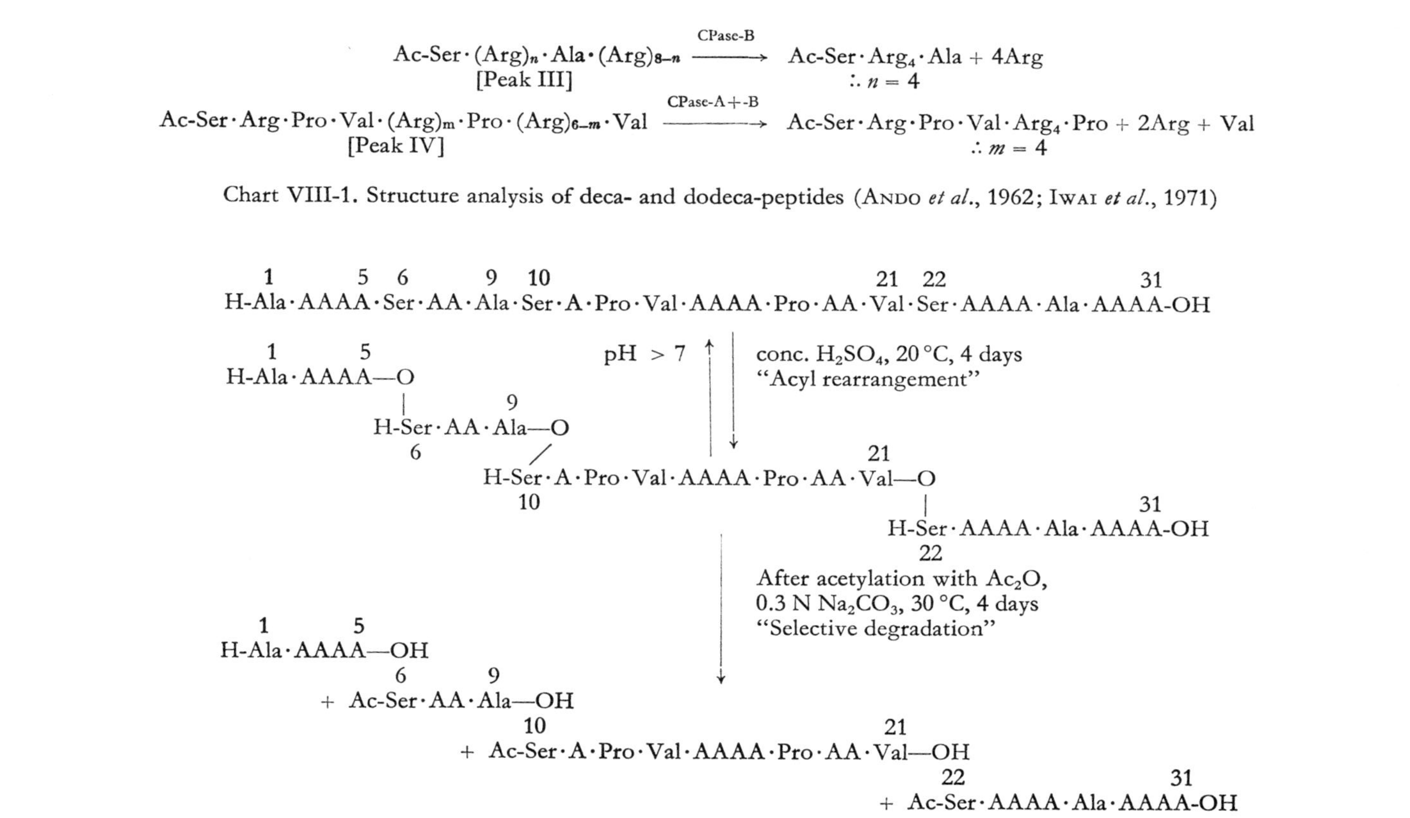

Chart VIII-1. Structure analysis of deca- and dodeca-peptides (ANDO *et al.*, 1962; IWAI *et al.*, 1971)

Chart VIII-3. N → O Acyl rearrangement reaction of clupeine Z followed by selective chemical degradation (ANDO *et al.*, 1962; IWAI *et al.*, 1971)

VIII-3) and carboxypeptidase B or A+ B (Fig. VIII-4) upon the protamine (ANDO *et al.*, 1962; AZEGAMI *et al.*, 1970). In these digestion studies, released amino acids were determined and the core peptides analysed. Modes of digestion are illustrated in Chart VIII-2.

It is concluded on the basis of the chemical structure determined for the protamine that the N → O acyl rearrangement reaction of clupeine Z and the selective chemical degradation proceed as shown in Chart VIII-3.

The clupeine Z molecule thus proved to have a peculiar amino-acid sequence which would not have been expected from the simple repeating structures hitherto

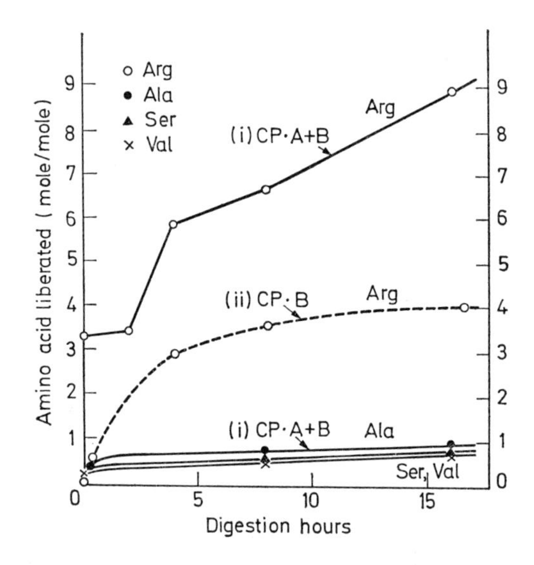

Fig. VIII-4. Hydrolysis of clupeine Z by carboxypeptidase (CP) B or A + B (from ANDO, 1964; AZEGAMI *et al.*, 1970)

suggested for clupeine [KOSSEL, 1928/1929; WALDSCMHIDT-LEITZ and KOFRANYI, 1935; FELIX *et al.*, 1952 (3); FELIX, 1960]. The structure is composed of such sequences as M·A·A, A·M·A, A·M·M·A and A·M·M·A·M·M·A (M represents mono-amino acid; A, arginine), as suggested earlier from the results of studies on the tryptic peptides of unfractionated or whole clupeine (ANDO *et al.*, 1959; ISHII *et al.*, 1967).

During enzymatic hydrolysis the possibility of transpeptidation (WALEY and WATSON, 1951; LEVIN *et al.*, 1956) should be borne in mind, especially in such a prolonged (20 h) reaction with trypsin. It seems unlikely, however, that the peptide in question, Ala·Ser·Arg·Pro·Val·Arg, was an artifact formed by transpeptidation. The peptide bond between arginine and proline is known to be quite resistant to tryptic hydrolysis, and the peptide is hydrolyzed slowly, but increasingly with time, into almost equal amounts of Ala·Ser·Arg and Pro·Val·Arg (0.2 μmole each after 88 min of tryptic hydrolysis, and 2.1 μmoles each after 20 h of hydrolysis from 11 μmoles of clupeine), as was discussed by ISHII *et al.* (1967). Some of the di- or tri-arginine residues found in the digest might, however, have been formed by transpeptidation, since lysyllysine was reported to be formed by transpeptidation in a prolonged tryptic digest of lysylglycineamide (LEVIN *et al.*, 1956).

B. The Amino-Acid Sequence of Clupeine YII

The amino-acid sequence of clupeine YII was determined in 1966 [ANDO and SUZUKI, 1966; SUZUKI and ANDO, 1972 (1)].

A rechromatographed specimen of clupeine YII obtained unmodified[1] by the chromatographic fractionation of TNP-(trinitrophenylated) whole clupeine (cf. Chap. VII. B. 3), has the molecular formula Arg_{20} Pro_3 Ala_2 Ser_2 Thr_1 Val_2. This is based on the analysis of amino-acid composition and a molecular weight of about 5,000 for the hydrochloride as estimated by the DNP method (MW calculated from the amino acid composition: 4,047 as a free base or 4,777 as the hydrochloride).

By the DNP and PTC (phenylthiocarbamyl) methods, the N-terminal amino acid was determined to be proline and the N-terminal amino acid sequence to be H-Pro·Arg·Arg···. The alternate action of carboxypeptidases B and A on clupeine YII followed by hydrazinolysis indicated that the amino-acid sequence at the C-terminal region was ···Pro·Arg_2·(Val, Ser)·Arg_4·Ala·Arg_4-OH.

A procedure essentially the same as that employed for clupeine Z (ANDO *et al.*, 1962; AZEGAMI *et al.*, 1970), enabled eleven peptides and arginine to be separated, identified and determined nearly quantitatively in a tryptic digest of clupeine YII. These tryptic peptides with the recovery (mole/mole) in parentheses are as follows: Thr·Arg (0.83), Val·Ser·Arg (0.87), Pro·Val·Arg (0.19), Ala·Ser·Arg (0.26), Ala·Arg (0.57 + 0.34), Pro·Arg (0.88), Arg (1.99), Ala·Ser·Arg·Pro·Val·Arg (0.64), Arg·Pro·Arg (1.03)[2], Arg.Arg (4.03), Arg·Arg·Pro·Arg (0.41), and Arg·Arg·Arg (0.25).

From the results described for both tryptic peptides and N- and C-terminal sequences, the following partial structure was derived for clupeine YII:
H-Pro·Arg_2·(Arg·Thr·Arg, Arg·Ala·Ser·Arg·Pro·Val·Arg, Arg)·Arg_2·Pro-Arg_2·Val·Ser·Arg_4·Ala·Arg_4-OH.

Larger peptide fragments retaining intact sequences of arginine residues were then obtained as with clupeine Z (ANDO *et al.*, 1962; IWAI *et al.*, 1971) by applying the N → O acyl rearrangement reaction at the serine and threonine residues in a clupeine YII molecule and selective chemical cleavage of the rearranged chain with weak alkali after acetylation. The degradation products were fractionated by gradient-elution chromatography on a CM-cellulose column to yield 3 main and a few minor fractions. The structures of these fractions were established, including numbers of intact arginine residues, by analysis of amino-acid composition and C-terminal residues; most of the amino-acid sequences from the C-terminals were determined by stepwise digestion with carboxypeptidases A and B, followed by hydrazinolysis when necessary, and subsequent analysis of the composition of the digestion cores. The yields (mole/mole) of acetylpeptides were as follows:

Ac-Thr·Arg·Arg·Ala (0.09), Ac-Pro·Arg·Arg·Arg (N-terminal, 0.16), Ac-Pro·Arg-Arg·Arg·Thr·Arg·Arg·Ala (N-terminal, 0.73), Ac-Ser·Arg·Pro·Val·Arg·Arg·Arg·Arg-Pro·Arg·Arg·Val (0.86), Ac-Ser·Arg·Arg·Arg·Arg·Ala·Arg·Arg·Arg·Arg (C-terminal, 0.91).

Chart VIII-4 sets out the complete amino-acid sequence of clupeine YII as concluded from these results, combined with the information derived from the tryptic peptides.

[1] Clupeine YII is not trinitrophenylated because it has proline at the N-terminus.

[2] Small amounts of Arg·Ala·Arg, Arg·Thr·Arg, and Arg·Val·Ser·Arg are also present in this peak; the recovery represents the total peak.

The structure of clupeine YII thus proved to be quite similar to that of clupeine Z, except for the substitution of the 6 residues at the N-terminal region (H-Ala·Arg_4-Ser-) in the Z-chain by the 5 (H-Pro·Arg_3·Thr-) in the YII-chain.

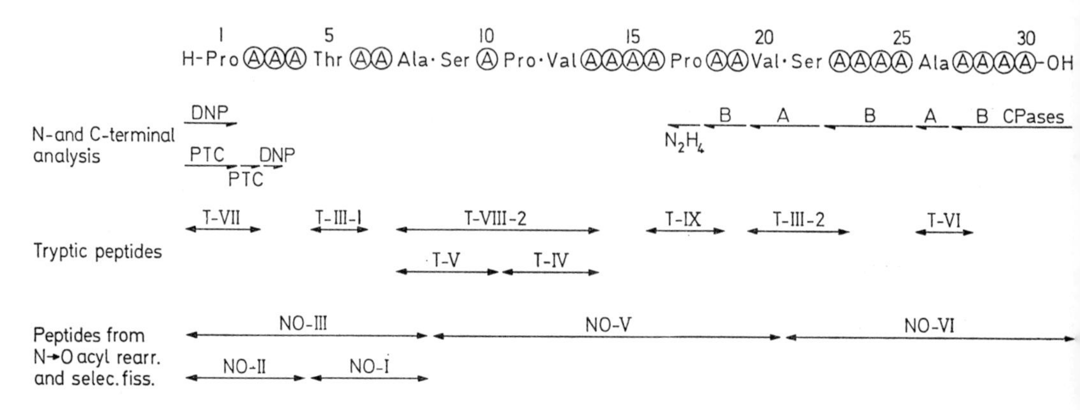

Chart VIII-4. The amino-acid sequence of clupeine YII with demonstration of amino acids obtained by terminal analyses, and peptides by tryptic hydrolysis and selective chemical hydrolysis after N → O acyl rearrangement reaction [Ando and Suzuki, 1966; Suzuki and Ando, 1972 (1)]

C. The Amino-Acid Sequence of Clupeine YI

The primary structure of the last component of clupeine, YI, was also determined by the Japanese workers, using a new method of enzymic hydrolysis with thermolysin in place of the N → O acyl shift and selective chemical degradation [Ando and Suzuki, 1967; Suzuki and Ando, 1972 (2)].

A rechromatographed specimen of clupeine YI was obtained as its TNP derivative by chromatography on CM-cellulose of TNP-whole clupeine or the TNP-Y fraction of clupeine [Ando and Suzuki, 1967; Suzuki and Ando, 1968 (2); Suzuki and Ando, 1972 (2)]. The molecular formula Arg_{20} Pro_2 Ala_2 Ser_3 Thr_2 Ile_1 Gly_1 was derived from the amino-acid analysis[3] and the molecular weight of approximately 5,000 as hydrochloride which was estimated from the absorption of the TNP-group (MW calculated from the amino-acid composition: 4,841 as hydrochloride and 4,112 as free base). From the enzymic digestion of TNP-clupeine YI described below, it was confirmed that in every case alanine was the N-terminal amino acid in this protamine molecule.

The sequence of sixteen amino-acid residues in the C-terminal region of TNP-clupeine YI forms more than half of the total chain; this was determined by the alternate action of carboxypeptidases B and A, followed finally by hydrazinolysis.

[3] During acid hydrolysis, TNP-amino acids liberate free amino acids in high yield (Okuyama and Satake, 1960). In the present case of TNP-alanine, free alanine was recovered in a yield of about 70%.

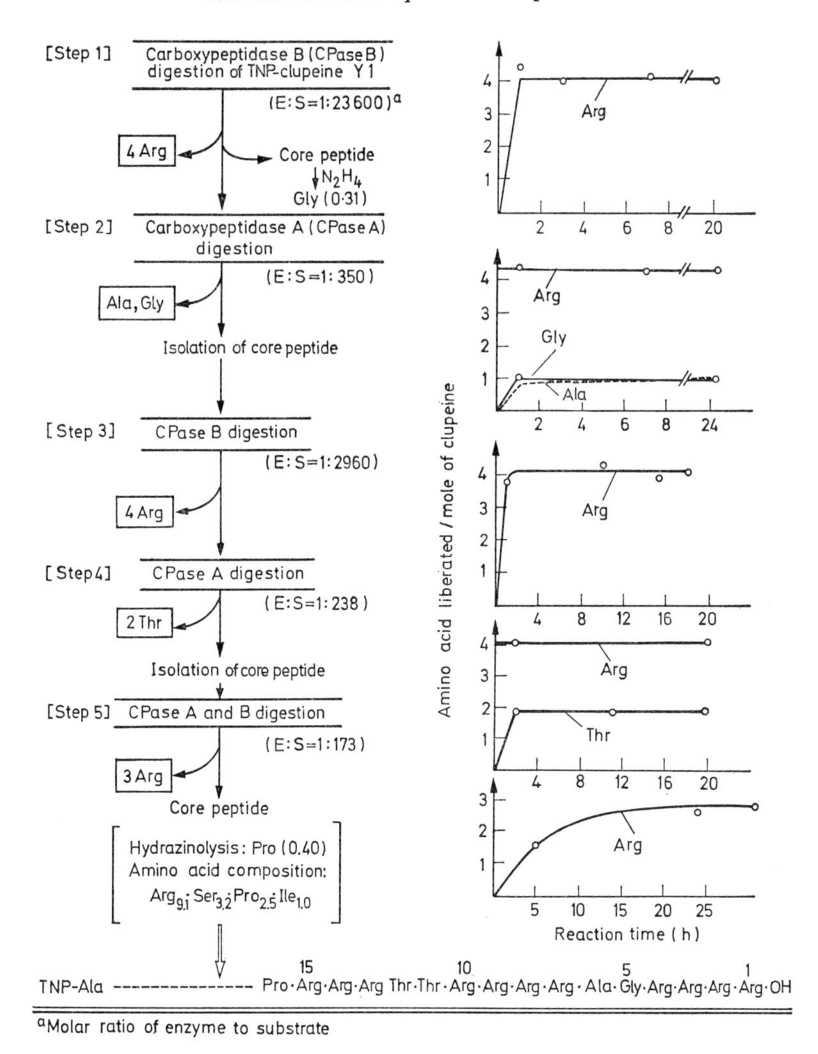

Fig. VIII-5. Schematic diagram and time course of liberation of amino acids from the C-terminal region of TNP-clupeine YI by the alternate action of carboxypeptidases B and A [from Suzuki and Ando, 1972 (2)]

These results are shown schematically in Fig. VIII-5, together with the time course of amino-acid liberation. Thus the amino-acid sequence at the C-terminal region was inferred as follows:

TNP-Ala .... Pro·Arg_3·Thr·Thr·Arg_4·Ala·Gly·Arg_4-OH.

The complete amino-acid sequence was established by the use of two complementary degradative procedures, hydrolysis by trypsin and thermolysin.

The tryptic peptides of TNP-clupeine YI were fractionated on an Amberlite CG-50 column and ten peptides and arginine were identified by employing essentially the same method as previously described (Ando *et al.*, 1959, 1962; Ando and Suzuki, 1966; Ishii *et al.*, 1967). These tryptic peptides with the recovery (mole/mole) in parentheses are as follows:

Thr·Thr·Arg (0.62), Ala·Gly·Arg (0.88), Ala·Arg (0.18), Pro·Arg (0.06, involved in a TNP-Ala·Arg peak), TNP-Ala·Arg (0.56), Arg (2.44), Ser·Ser·Ser·Arg·Pro·Ile·Arg (0.76), Arg·Thr·Thr·Arg (0.32), Arg·Pro·Arg (0.22), Arg·Arg (5.82), Arg·Arg·Arg (undetermined), Arg·(Arg, Pro)·Arg (ca. 0.25).

From the results with tryptic peptides and C-terminal sequence analysis, the following partial structure is formulated tentatively for TNP-clupeine YI:

$$\text{TNP-Ala}\cdot\text{Arg}\cdot(\text{Ser}_3.\text{Arg}\cdot\text{Pro}.\text{Ile}\cdot\text{Arg, 5 Arg})\cdot\text{Arg-}$$
$$\text{Pro}\cdot\text{Arg}_3\cdot\text{Thr}_2\cdot\text{Arg}_4\cdot\text{Ala}\cdot\text{Gly}\cdot\text{Arg}_4\text{-OH}$$

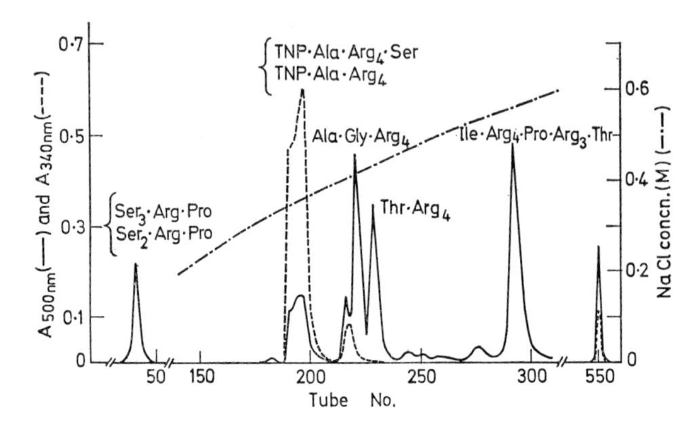

Fig. VIII-6. Chromatographic separation and identification of peptides in a thermolysin digest of TNP-clupeine YI. The digestion mixture (from 2.27 μmoles of TNP-clupeine YI) was applied to a column (0.9 cm × 150 cm) of CM-cellulose. Peptides were first eluted with the starting buffer (0.1 M acetate buffer, pH 6.0; tubes 1—80), then an exponential gradient elution was applied (81—514). The mixing chamber contained 1000 ml of starting buffer and to the reservoir 0.1 M acetate buffer (pH 6.5) containing 1.0 M NaCl was added. The peptides remaining on the column were eluted with 0.1 M HCl. The flow rate of the effluent was 3.89 ml/tube/18 min. The effluent was analyzed by SAKAGUCHI colorimetry and spectrophotometry at 340 mμ. The color yield (SAKAGUCHI reaction) was 101% (tubes 1 to 555) [from ANDO and SUZUKI, 1967; SUZUKI and ANDO 1972 (2)]

In the case of TNP-clupeine YI, neither selective chemical cleavage of the N → O acyl rearranged product nor chymotryptic digestion produced peptide fragments retaining intact sequences of arginine residues. The former procedure, which was used with success for clupeine fractions YII and Z, gave a low recovery of many fragments which were hard to separate into homogeneous parts. The presence of groups of serine and threonine residues (-Ser·Ser·Ser- and -Thr·Thr-) in the clupeine YI molecule might be one of the reasons for such a complication.

In place of selective chemical degradation, enzymic digestion using a thermostable bacterial protease, thermolysin, obtained from *Bacillus thermoproteolyticus* Rokko (ENDO, 1962; MATSUBARA *et al.*, 1965), was successfully applied to clupeine YI [ANDO and SUZUKI, 1967; SUZUKI and ANDO, 1972 (2)].

The digestion was performed in 0.01 M calcium chloride at 40 °C for 3.5 h with a pH-stat at pH 8.0, in a molar ratio of substrate to enzyme of 119:1. Peptides in the digestion mixture were fractionated by gradient-elution column chromatography on CM-cellulose (Fig. VIII-6). The amino-acid sequences of the peptides in the five main peaks were established as usual by determination of amino-acid composition, by use of the DNP and PTC methods, by stepwise digestion with carboxypeptidases A and B, followed finally by hydrazinolysis. These five thermolysin peptides completely accounted for the composition of TNP-clupeine YI. The recovery values (mole/mole) of thermolysin peptides were as follows:

Ser_3·Arg·Pro and Ser_2·Arg·Pro (0.98, in total), TNP-Ala·Arg_4·Ser and TNP-Ala·Arg_4 (0.54, in total), Ala·Gly·Arg_4 (0.77), Thr·Arg_4 (0.82 , and Ile·Arg_4·Pro·Arg_3·Thr (0.73).

These figures show that thermolysin hydrolyzes no arginylarginine bonds at all, and furthermore, no bonds between basic amino acids (Hayashi, Iwai, and Ando, unpublished data). Thermolysin is thus an excellent new tool for sequence deter-

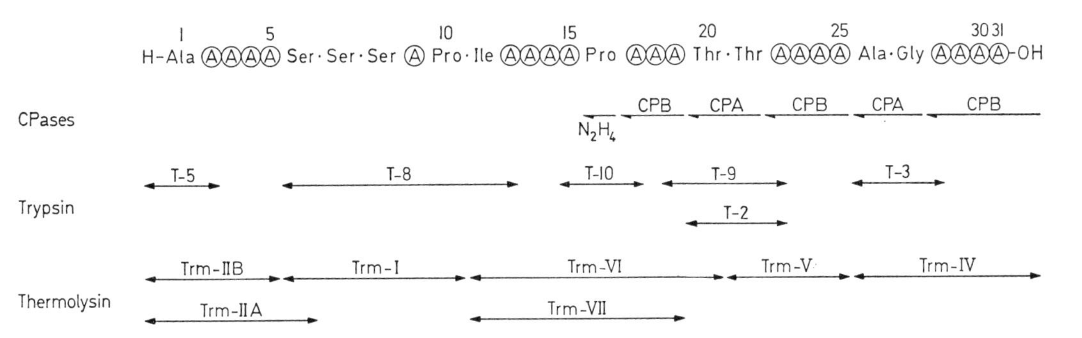

Chart VIII-5. The amino-acid sequence of clupeine YI with demonstration of peptides obtained by hydrolysis with carboxypeptidases B and A, trypsin and thermolysin [Ando and Suzuki, 1967; Suzuki and Ando, 1972 (2)]

mination in protamines, and probably in histones and other basic proteins, as well as for obtaining blocks of basic amino acids.

The existence of two sets of overlapping peptides resulting from digestion with trypsin and thermolysin, combined with the results of the action of carboxypeptidases B and A used alternately, made it possible to deduce the total amino-acid sequence of clupeine YI as shown in Chart VIII-5.

D. Comments on the Structure of Clupeine

Some consideration has been given to the three components of clupeine [Ando and Suzuki, 1967; Suzuki and Ando, 1972 (1, 2)].

The primary and secondary structures of all the molecular species of clupeine (from Pacific herring, *Clupea pallasii*), YI, YII and Z, have been elucidated [see Chart VIII-6 for the primary structures, and Suzuki and Ando, 1968 (1), for the secondary structures]. This is the first and only instance at present of structure determination and elucidation of the number of molecular species present in the

```
        1      5                        10          15          20          25           31
YI:   H-Ala·AAAA·        Ser·Ser·Ser·A·Pro·Ile ·AAAA·Pro·AAA·Thr·Thr·AAAA·Ala·Gly·AAAA-OH

        1                                                                                30
YII:  H-Pro·AAA • Thr·AA·Ala·Ser·A ·Pro·Val·AAAA·Pro·AA  ·Val·Ser ·AAAA ·Ala·        AAAA-OH

        1      5              10           15        20            25                  31
Z:    H-Ala·AAAA·Ser·AA·Ala·Ser·A ·Pro·Val·AAAA·Pro·AA  ·Val·Ser ·AAAA·Ala·         AAAA-OH
```

Chart VIII-6. The total primary structures of 3 main components of clupeine (A = Arg)

basic nuclear proteins, histones and protamines. At the same time the methods of approach to the primary structure of basic proteins, especially protamines, have been generally established.

The amino-acid sequence of clupeine YI, though somewhat different from that of the other two species (YII and Z), bears an overall structural resemblance to the others. The most striking similarity is that all of the components contain near the N-terminal hepta- (Ser_3·Arg·Pro·Ile·Arg) or hexa- (Ala·Ser·Arg·Pro·Val·Arg) peptide rich in neutral amino acids, succeeded by the arginine-rich sequence (-Arg. Arg·Arg·Pro·Arg.Arg-). The three molecules also have a similar distribution of hydrophobic and hydroxyl amino acids and proline residues. On the other hand, there is a marked lack of regularity within each structure rather than the repeating structure suggested by Felix *et al.* (1960). The peptide chains of the three molecular species of clupeine contain several parts where a small number of single neutral amino acids separate arginine blocks (proline in all three, alanine and threonine in YII, and alanine and serine in Z). Since none of these amino acids can form loops in combination with DNA, some phosphate groups of DNA will remain uncombined with arginine residues. Hence the model of the DNA-clupeine complex proposed by Wilkins (1956) based on the hypothetical repeating structure of protamine must be changed, at least in this respect.

Neither the exhaustive studies of the Tokyo group (1964 and 1969) as described in the next section (Chap. VIII. E) nor the partial study by Felix and Hashimoto (1963) have revealed any differences in the amino-acid sequences of the corresponding components of clupeine from Norwegian Sea and North Sea *(Clupea harengus)* and Pacific *(Clupea pallasii)* herring.

Evidence has recently been accumulating that histone, which is present in somatic cell nuclei, regulates or modifies gene activity (Stedman and Stedman, 1950, 1951; Huang and Bonner, 1962; Bonner and Huang, 1964; Busch, 1965; Allfrey, 1966). Although the genetic function of protamines (see also Chap. X. C) is still obscure, they do have some role, if we accept that DNA-primed RNA synthesis is a direct measure of genetic function (Skalka *et al.*, 1966; Suzuki and Ando, 1969).

Studies on the secondary structure of clupeine [Suzuki and Ando, 1968 (1); see also Chap. IX. B] indicate that the three component molecules have no helical parts in their chains and that their conformation in physiological condition is completely random. Therefore, not only their primary structures but also their secondary structures must be essentially the same. The significance of the presence of these three similar molecules in clupeine and the question whether each plays a different and specific role in biological activity are matters of fundamental interest which await further study.

From the standpoint of evolution, Black and Dixon (1967) proposed a hypothesis concerning the evolutionary pathway of the components of clupeine. They employed a computer program based on the theory of partial gene duplication. They supposed that each of the clupeine molecules has evolved from an ancestral pentapeptide, Ala·Arg·Arg·Arg·Arg, by repeated partial gene duplication in which the insertion, deletion and replacement of amino acids were effected by single changes of bases within the coding triplets of DNA. On the basis of the number of mutational changes, the clupeine Z and YI sequences were thought to have diverged initially, and the Z and YII sequences more recently. A model proposed by Black and Dixon for the evolution of clupeine Z from an ancestral pentapeptide is shown below in Chart VIII-7. More recently, FITCH (1971) proposed an alternative scheme for history of clupeines assuming Ala·Arg_5 as an archetypal hexapeptide and clupeine Z to be a crossover product.

E. The Amino-Acid Sequences of the Three Components (Y′I, Y′II and Z′) of Clupeine from North Sea Herring

Clupeine obtained from sperm cell nuclei of *Clupea harengus* has been studied for about 50 years, since the time of KOSSEL, by European protein researchers, and facts have accumulated on some features of the chemical structure of unfractionated (whole) clupeine. Over 10 years ago, some differences were noted in the amino-acid composition and the N-terminal amino acid residue of clupeine specimens prepared from Norwegian Sea (FELIX *et al.*, 1950; WALDSCHMIDT-LEITZ *et al.*, 1951) and Pacific [ANDO *et al.*, 1952, 1953, 1957 (1, 2)] herring. It was considered by both the German and the Japanese workers that such differences were due to the species specificity of the materials used.

However, all these results were obtained by analysis of unfractionated clupeine. The complete amino-acid sequences recently determined for the three components of clupeine from Pacific herring *(Clupea pallasii)* (ANDO *et al.*, 1962; ANDO and SUZUKI, 1966, 1967) confirmed the presence of an amino-acid residue, glycine, as a constituent and of the N-terminal alanine in addition to the N-terminal proline in some of the clupeine molecules. At that time no complete fractionation of clupeine from *Clupea harengus* into a number of homogeneous components had been accomplished.

In 1963, FELIX and HASHIMOTO revised their earlier results on the N-terminals and the amino-acid composition of unfractionated clupeine from *Clupea harengus*, so that there was no longer any essential difference in material from the two herring species. A sample of unfractionated clupeine from North Sea herring sperm heads was prepared again in the late Prof. FELIX's laboratory in Frankfurt am Main in March 1963 and sent to Prof. ANDO's laboratory in Tokyo for use in structural studies. This sample was found to be not different from unfractionated clupeine from Pacific herring as regards amino-acid composition, N-terminal amino acids (proline, 36% and alanine, 64%), N-terminal sequences (Pro·Arg··· and Ala·Arg·Arg···), and tryptic peptides (Thr·Thr 0.95, Thr·Arg 0.84, Val·Ser·Arg 1.7, Pro·Val·Arg 0.3_3, Pro·Ile·Arg 0.1_2, Ala·Gly·Arg 1.0_3, Ser·Arg 0.52, Ala·Ser·Arg 0.27, Ala·Arg 3.5_4, Ser·Ser·Ser·Arg·Pro·Ile·Arg 0.7_8, Pro·Arg 0.8_7, Ala·Ser·Arg·Pro·Val·Arg 1.0_2, Arg 8.16, Arg·Pro·Arg 1.5, Ala·Arg·Arg 0.5, Arg·Arg 10.2 moles/3 moles clupeine) (NUKUSHINA, 1964; NUKUSHINA *et al.*, 1964).

More recently the Japanese group (CHANG, 1969; CHANG, NAKAHARA and ANDO, to be published) fractionated clupeine from North Sea herring into the components clupeine Y′I, Y′II and Z′, by the method described in Chap. VII. B. 4 for the fractionation of clupeine from Pacific herring (ANDO and WATANABE, 1969). The complete amino-acid sequences of the components Y'I, Y'II and Z' were obtained from the results of N- and C-terminal sequence analysis and analysis of the thermolytic peptides, and from data on the structure of the tryptic peptides of unfractionated clupeine from *Clupea harengus* (see p. 37; ref. 7). They are illustrated in Chart VIII-8.

There is hardly any difference in the primary structure found for the corresponding components of clupeine from North Sea and Pacific herring. It is thus concluded that

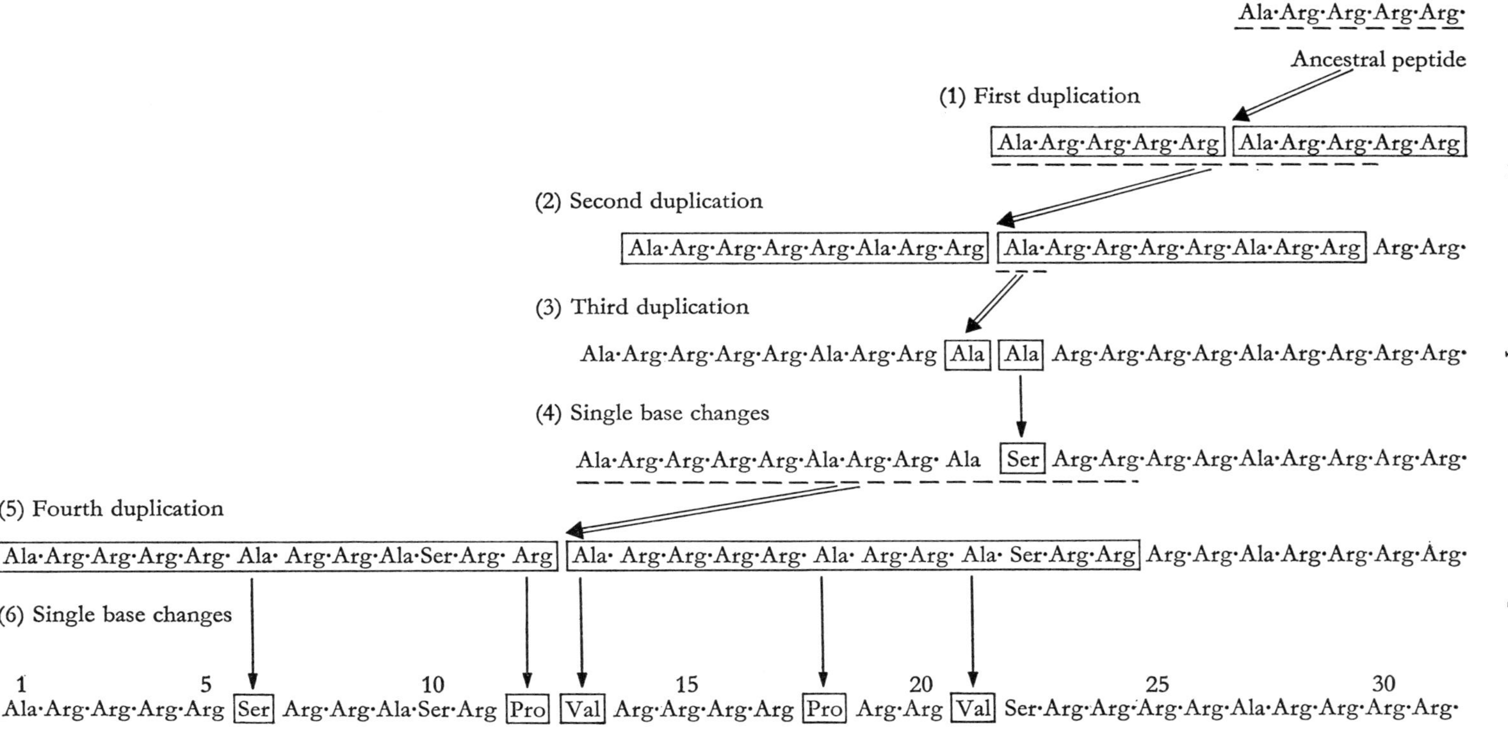

Chart VIII-7. A proposed model for the evolution of clupeine Z from an archetypal pentapeptide by a series of partial gene duplications and single base changes in the structural gene for this protein. The segment of polypeptide underlined with a dotted line corresponds to that portion of the structural gene which is partially duplicated and the resulting new polypeptide sequence is enclosed in a box. Single arrows represent single base changes while the double arrows show the partial gene duplications (from BLACK and DIXON, 1967)

there is no species specificity in the amino-acid sequences of clupeine from the two species in this genus. This meant that the views held previously by both the German and the Japanese researchers had to be revised.

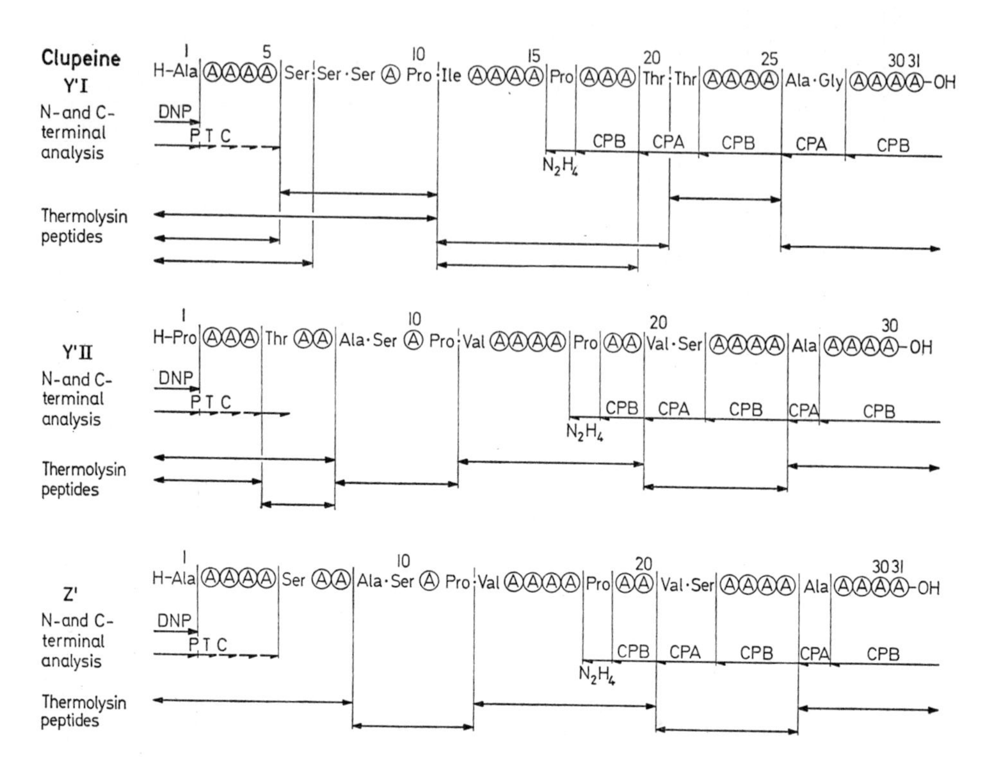

Chart VIII-8. The complete amino-acid sequences deduced for clupeine Y'I, Y'II and Z' with illustration of several degradations. (Structures of tryptic peptides obtained from whole clupeine from *Clupea harengus* were also referred to for this deduction. See text) (A = Arg) (CHANG, 1969; CHANG, NAKAHARA, and ANDO, to be published)

F. The Amino-Acid Sequences of One Component of Salmine and Three Components of Iridine

The amino-acid sequences of some components obtained homogeneously from salmine and iridine have recently been established with the help of a new enzymic degradation method in addition to the usual procedures (ANDO and WATANABE, 1969; WATANABE and ANDO, to be published).

A rechromatopgraphed specimen of one component of salmine (S-AI) obtained by chromatography on a column of Bio-Gel CM-2 or CM-Sephadex C-25 (cf. Chap. VII. B. 4) of whole salmine from Hokkaido salmon *(Oncorhynchus keta)* [ANDO *et al.*, 1957 (1, 3); YAMASAKI, 1958] has the molecular formula Arg_{21} Ser_4 Pro_3 Gly_2 Val_2. Rechromatographed specimens of two components of iridine (I-I and I-II), obtained by the same method from whole iridine from the rainbow trout *(Salmo irideus)*, have the molecular formula Arg_{22} Ser_4 Pro_3 Gly_2 $Val_{1.3}$ $Ile_{0.7}$ and Arg_{21} Ser_4 Pro_2 Gly_2 Ala_1 Val_2, respectively. The first component of iridine, I-I, was later found to consist of two components, I-Ia and I-Ib. These were not separated by chromatography because their amino-acid composition is the same except for the presence of one isoleucine residue in I-Ib in place of one valine in I-Ia, as determined by structural analysis of peptides obtained on digestion with thermolysin and Neutral Protease (see below).

The structural studies of these components were performed in essentially the same way as described for clupeine YI [ANDO and SUZUKI, 1967; SUZUKI and ANDO, 1972 (2)]. The N-terminus of each of the three components (S-AI, I-I and I-II) was determined by the DNP method to be proline. The use of carboxypeptidase B followed by hydrazinolysis, or the alternate use of carboxypeptidases B and A, suggested that the amino-acid sequence of all three components has, in common, more than four arginine residues followed by approximately two glycine residues from the C-terminus ($\cdots Gly_2 \cdot Arg_4$-OH). The results of the digestion further suggested that arginine, serine, and valine are present at the next C-terminal regions in salmine (S-AI) and iridine (I-I) components, and arginine, serine, and valine, plus alanine, in the other iridine component (I-II).

Table VIII-5. Peptides obtained by thermolysin digestion of iridine I[a]

Structure	Recovery (mole/mole)	Thought to be derived from
$Pro \cdot Arg_m \cdot Ser \cdot Ser \cdot Ser \cdot Arg \cdot Pro \cdot Val \cdot Arg_n \cdot Pro \cdot Arg_2$ ($m + n = 9$)	0.17	Component (a) (I-Ia)
$Val \cdot Ser \cdot Arg_6 \cdot Gly \cdot Gly \cdot Arg_4$	0.18	
$Pro \cdot Arg_6 \cdot Ser \cdot Ser \cdot Ser \cdot Arg \cdot Pro$	0.16	Component (b) (I-Ib)
$Pro \cdot Arg_6 \cdot Ser \cdot Ser \cdot Ser \cdot Arg \cdot Pro \cdot Ile \cdot Arg_4 \cdot Pro \cdot Arg_2$	0.47	
$Val \cdot Ser \cdot Arg_5 \cdot Gly \cdot Gly \cdot Arg_4$	0.61	

[a] These results revealed the presence of 2 molecular species in iridine I and also the different distribution of arginine clusters between the two (compare also with the results of the Neutral Protease peptides in Table VIII-6), where *m* and *n* are determined to be 4 and 5 respectively).

Tryptic peptides (1/15 M phosphate buffer, pH 8.0, 37 °C, 20—25 h, trypsin treated with diphenylcarbamyl chloride, substrate/enzyme = 100—250, w/w) were obtained in good yields as illustrated in the chart of the complete amino-acid sequences (cf. Chart VIII-9).

Fragmentation into larger peptides retaining intact sequences of arginine residues was achieved by hydrolysis with thermolysin (1/15 M phosphate buffer, pH 8.05, 37 °C, 3 h, substrate/enzyme = 200—300, w/w). The thermolysin peptides obtained with good recovery are also illustrated in Chart VIII-9. The structure and recovery of the thermolysin peptides obtained from iridine I are shown in Table VIII-5 as well as Chart VIII-9. The results revealed the presence of two molecular species (I-Ia and I-Ib) in the iridine fraction (I-I). Some difference between the two species of molecules was also found in the number of arginine residues in their clusters. This observation was further supported by the results of Neutral Protease peptides, as described below.

In order to obtain more precise information on the distribution of arginine residues among their blocks in these longer thermolytic peptide fragments, a new method of hydrolysis was sought which would give somewhat smaller peptides while retaining the arginine sequences intact. A method which proved successful was digestion of the protamines (salmine AI and iridine I) with a "Neutral Protease" from *Bacillus subtilis var. amylosacchariticus* (in 0.002 M calcium chloride, with a pH-stat at pH 7.2, 3 to 24 h at 37 °C, enzyme/substrate ratio 1:20 by weight). This enzyme was discovered, purified, and characterized in Japan (TSURU *et al.*, 1967). In the case of the second iridine

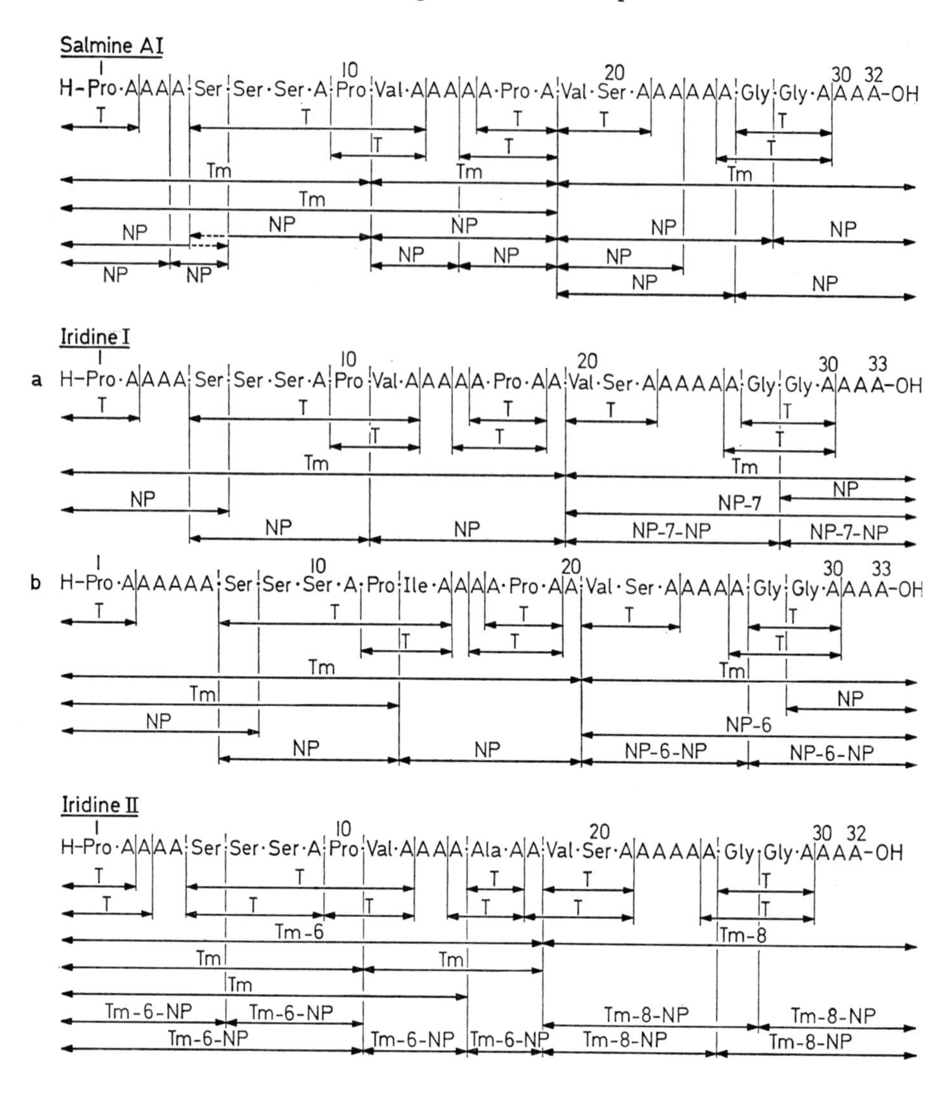

Chart VIII-9. The amino acid sequences of salmine AI, iridine I (a) and (b), and iridine II with demonstration of peptides obtained by enzymic hydrolyses. The symbols used are: A, Arg; T, tryptic peptide; Tm, thermolysin peptide; NP, "Neutral Protease" peptide; NP-7-NP, redigested "Neutral Protease" peptide from NP-No. 7; Tm-6-NP, "Neutral Protease" peptide from Tm-No. 6 (from ANDO and WATANABE, 1969)

component (I-II), two thermolysin peptides consisting of residues 1—18 (Tm-6 in Chart VIII-9) and 19-32 (Tm-8), but not the iridine component itself, were hydrolyzed by Neutral Protease. Peptides in the digestion mixture were fractionated satisfactorily by column chromatography on CM-Sephadex C-25 (0.9 × 150 cm, 0.05 M acetate buffer, pH 5.8, linear gradient elution of 0.5 → 1.5 M NaCl, for example).

Table VIII-6. Peptides obtained by Neutral Protease digestion

Salmine AI

Structure	Recovery (mole/mole)
Pro·Arg·Arg·Arg·Arg Pro·Arg·Arg·Arg·Arg·Ser	0.37
Arg·Ser	0.42
Pro·Arg·Arg·Arg	0.44
(Ser)·Ser·Ser·Arg·Pro	0.96
Val·Arg·Arg·Arg·Arg·Arg·Pro·Arg	0.80
Arg·Arg·Pro·Arg	0.12
Val·Arg·Arg·Arg	0.07
Val·Ser·Arg·Arg·Arg·Arg·Arg·Arg Val·Ser·Arg·Arg·Arg·Arg·Arg·Arg·Gly	0.83
Val·Ser·Arg·Arg·Arg	0.10
Gly·Gly·Arg·Arg·Arg·Arg	0.48
Gly·Arg·Arg·Arg·Arg	0.50

Iridine I[a]

Structure	Recovery (mole/mole)	Supposed to be derived from
Pro·Arg·Arg·Arg·Arg·Ser	0.17	component (a) (I-Ia)
Val·Arg·Arg·Arg·Arg·Arg·Pro·Arg·Arg	0.29	
Val·Ser·Arg·Arg·Arg·Arg·Arg·Arg·Gly·Gly·Arg·Arg·Arg·Arg	0.13	
Ser·Ser·Ser·Arg·Pro	0.47	both components (a) and (b)
Gly·Arg·Arg·Arg·Arg	0.17	
Pro·Arg·Arg·Arg·Arg·Arg·Arg·Ser	0.82	component (b) (I-Ib)
Ile·Arg·Arg·Arg·Arg·Pro·Arg·Arg	0.61	
Val·Ser·Arg·Arg·Arg·Arg·Arg·Gly·Gly·Arg·Arg·Arg·Arg	0.55	

[a] These Neutral Protease peptides from iridine I revealed, along with the thermolysin peptides, the presence of 2 molecular species of the same composition except for substitution of a valine residue with an isoleucine residue between the two.

Iridine II

From Neutral Protease digest of thermolysin peptide-6 (residues No. 1—18):	Recovery (mole/mole)
Pro·Arg·Arg·Arg·Arg·Ser	0.46
Pro·Arg·Arg·Arg·Arg·Ser·Ser·Ser·Arg·Pro	0.11
(Ser)·Ser·Ser·Arg·Pro	0.75
Val·Arg·Arg·Arg·Arg	0.39
Ala·Arg·Arg	0.28

From Neutral Protease digest of thermolysin peptide-8 (residues No. 19—32):	Recovery (mole/mole)
Val·Ser·Arg·Arg·Arg·Arg·Arg·Arg	0.85
Val·Ser·Arg·Arg·Arg·Arg·Arg·Arg·Gly	0.10
Gly·Gly·Arg·Arg·Arg·Arg	0.39
Gly·Arg·Arg·Arg·Arg	0.46

The chemical structures of these peptides were established in the usual way by quantitative amino-acid analysis and sequence analysis by a combination of chemical and enzymic methods, including dinitrophenylation, dansylation, Edman degradation, hydrazinolysis and digestion with leucine aminopeptidase and carboxypeptidases A and B. The Neutral Protease peptides of the salmine and iridine components or derivatives thus identified are listed in Table VIII-6 with their recovery values. Differences in the amino-acid sequences and amounts of the peptides obtained from iridine I aroused the suspicion that two molecular species (a and b) were present in the iridine I component. This result, as already described, was supported by observations on the thermolysin peptides of iridine I.

No arginylarginine bonds are hydrolyzed by Neutral Protease in a short digestion period, although hydrolysis takes place to a minor extent during prolonged digestion. The use of a bacterial protease, Neutral Protease, for digestion is thus a new method of degradation, complementary to the use of trypsin and thermolysin for the determination of protein structure. The method is especially useful for such basic proteins as protamines and histones, in which blocks of basic amino acids are expected to be present.

To summarize: for one component of salmine (S-AI) and three components of iridine (I-Ia, I-Ib and I-II) complete amino-acid sequences were deduced as shown in Chart VIII-9, from the information obtained from three sets of peptides resulting from digestion with trypsin, thermolysin, and Neutral Protease, as well as from the N- and C-terminal sequences.

Chapter IX

Physical Structure of Nucleoprotamines and Protamines

A. Nucleoprotamines

Native and reconstituted nucleoprotamines and sperm nucleus itself give essentially identical X-ray diffraction patterns which are also similar to the pattern of DNA (WILKINS *et al.*, 1953; WILKINS and RANDALL, 1953; FEUGHELMAN *et al.*, 1955; WILKINS, 1956). Infrared absorption spectra show that the DNA portion of DNA-protamine has a structure identical to that of the free DNA molecule [BRADBURY *et al.*, 1962 (1)]. Therefore DNA *in vivo* is presumably in the B form, whether or not protamine is bound to it. Judging from the hydrogen-deuterium exchange reaction [HAGGIS, 1957; BRADBURY *et al.*, 1962 (1)] and infrared absorption spectra, the structure of the protamine portion in the DNA-protamine complex is supposed to be a partly extended form, intermediate between the α-helix and fully extended β-structure [BRADBURY *et al.*, 1962 (1)]. Accordingly, in the nucleoprotamine molecule, B-form DNA and partly extended protamine are thought to combine in some way so that basic groups in protamine are bound to the phosphate groups of DNA (FEUGHELMAN *et al.*, 1955; WILKINS, 1956). The physical structure of nucleoprotamine has not yet been established, but three models have been proposed for it.

The first model (Fig. IX—1), which is based on X-ray analyses (FEUGHELMAN *et al.*, 1955; WILKINS, 1956; ZUBAY and WILKINS, 1962), shows protamine winding around the small groove on the DNA helix, with side-chains of arginine residues stretching out on alternate sides of the polypeptide chain to neutralize the phosphate groups of DNA. At the time this model was proposed, neutral amino acids were thought to be present in pairs among arginine blocks (FELIX, 1960) and looped out as shown in the figure. Although this model accounts well for the X-ray diffraction data, some facts have been found which are incompatible with it. For example, free DNA molecules can swell freely as the relative humidity (R.H.) increases; however, the distance between the two adjacent DNA molecules in nucleoprotamine is rather limited, the value for the main equatorial reflection being about 25 Å even at 100% R. H. (FEUGHELMAN *et al.*, 1955). This indicates the existence of some bridging force between two adjacent DNA molecules which limits their swelling (ZUBAY and DOTY, 1959). Furthermore, the amino-acid sequence analyses of clupeine and salmine indicate that neutral amino acids are present not only as doublets but also as singlet and triplet forms (ANDO *et al.*, 1962; ANDO and SUZUKI, 1966, 1967). Since it is impossible to loop out neutral amino acids which exist in singlet form, the model needs some correction, in this respect at least.

The second model of nucleoprotamine, proposed by LUZZATI and NICOLAIEFF (1963) and LUZZATI (1963), consists of hexagonally arranged DNA molecules with protamine and water filling up the gaps, so that it has the structure of an infinite

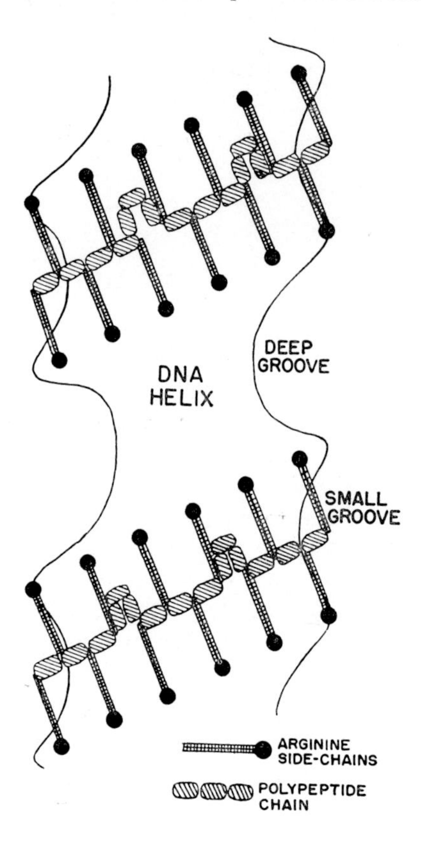

Fig. IX-1. Diagram showing how protamine binds to DNA. The polypeptide chain winds around the small groove on the DNA helix. The phosphate groups are at the black circles and coincide with the basic ends of the arginine side-chains. Non-basic residues are shown in pairs at folds in the polypeptide chain (from WILKINS, 1956)

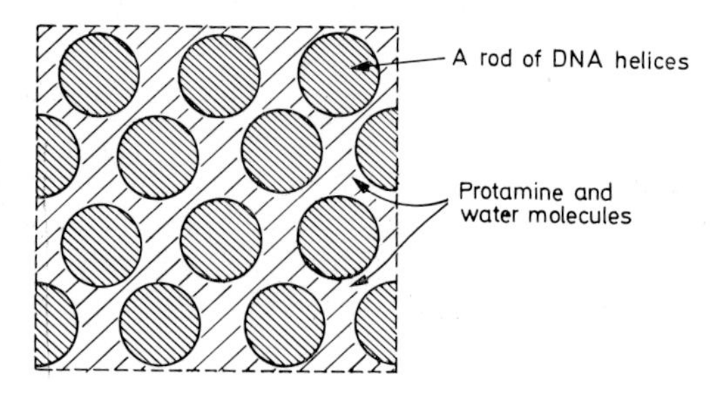

Fig. IX-2. A nucleoprotamine model (shown in section perpendicular to the rod direction) consisting of two-dimensional hexagonal array of DNA molecules, with protamine and water molecules filling up the gaps (from LUZZATI and NICOLAIEFF, 1963)

network (Fig. IX-2). In this model a single "nucleoprotamine molecule" cannot exist, whereas in the first model it can. This model can explain the data on limited swelling, but not the fact that nucleoprotamine can be extracted with 0.075 M NaCl solution containing 0.024 M EDTA at pH 8 (ZUBAY and WILKINS, 1962).

This result was taken into consideration in the third model (Fig. IX-3), based on the data from small-angle X-ray diffraction and electron-microscope studies [RAUKAS *et al.*, 1966 (1, 2)]. In this case four DNA-protamine molecules which are very similar to those in the first model are presumed to be bound by protamine and/or metal ions (such as Mg^{++} or Ca^{+}).

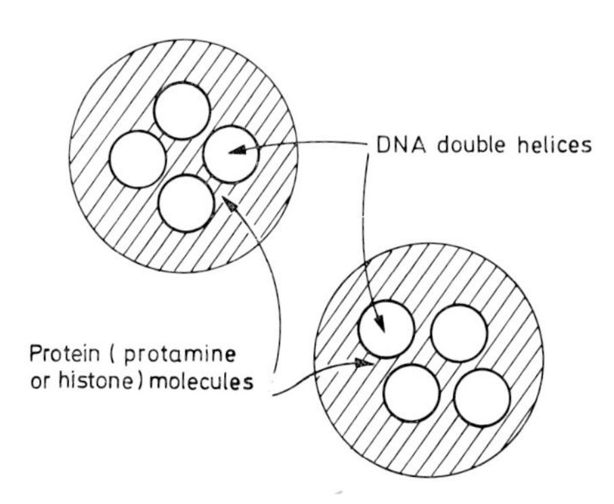

Fig. IX-3. A nucleoprotamine model (shown in section perpendicular to the rod direction) each unit consisting of four rods of DNA double helices joined together in places probably by protamine bridges or divalent metallic ions [from RAUKAS *et al.*, 1966 (1, 2)]

Recently the Japanese group investigated the interaction of clupeine or basic oligopeptides with DNA using techniques of melting profile (INOUE and ANDO, 1966; KAWASHIMA *et al.*, 1969; see also Chap. X. C. 1), sedimentation [INOUE and ANDO, 1970 (1)], optical rotatory dispersion (INOUE and ANDO, 1968, 1969), gel filtration (INOUE and ANDO, 1969), and electron microscopy (INOUE and FUKE, 1970; see Fig. IX-4). On the basis of these results, they proposed a model as shown in Fig. IX-5 for reconstituted DNA-clupeine.

The structure of nucleohistone appears to be rather different from that of nucleoprotamine, since its X-ray diffraction pattern is different (WILKINS and ZUBAY, 1959; ZUBAY and WILKINS, 1962; PARDON *et al.*, 1967), also the position of the amide II band in the infrared absorption spectrum and the rate of deuteration [BRADBURY *et al.*, 1962 (2)].

B. Protamines

The deuterium exchange reaction of protamine is very fast, less than 3 min being required for the complete deuteration of a film exposed to D_2O vapour [LINDERSTRØM-LANG, 1956; BRADBURY *et al.*, 1962 (1)]. This indicates that protamine is not in α-helical configuration but in some extended form. Protamine has the amide I band at 1650 to 1660 cm^{-1} and the amide II band at 1552 cm^{-1}, which is characteristic

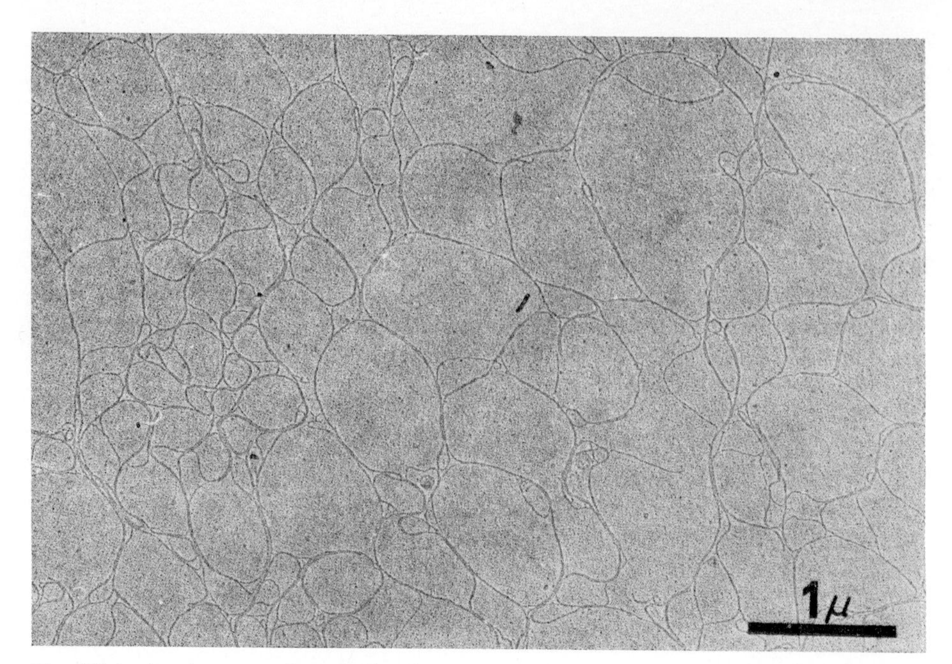

Fig. IX-4. An electron micrograph of reconstituted DNA-clupeine complexes. The complexes, formed by the slow-binding method from native herring sperm DNA and whole clupeine, were dissolved in 4 M ammonium acetate. Samples for electron microscopy were prepared according to the method of KLEINSCHMIDT *et al.* (1962) and shadow-casted with platinum-palladium (80:20, v/v) (from INOUE and FUKE, 1970)

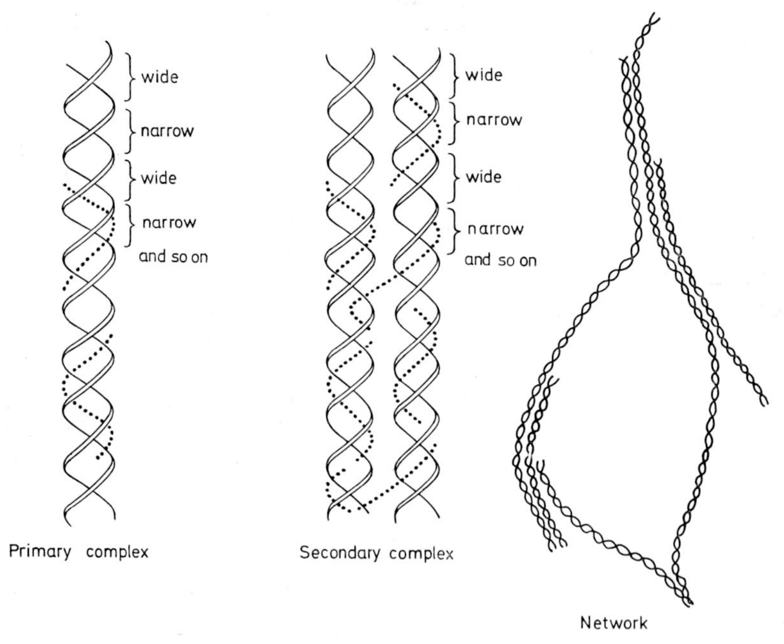

Fig. IX-5. Schematic models of DNA-clupeine complexes based on the thermal denaturation, optical rotatory dispersion, gel-filtration, and electron microscope studies. Primary complex: clupeine molecules wind around the small grooves on the DNA double helix. Secondary complex: the primary complexes are joined together in places by protamine bridges. Network: a network of nucleoprotamine complexes as can be seen in the electron micrograph (cf. Fig. IX-4) (from INOUE and ANDO, 1969)

of proteins in either the α-helical or the random-coil form. Hence protamine must be in the random-coil structure [BRADBURY *et al.*, 1962 (1)].

Analysis of protamine by optical rotatory dispersion shows that its b_0 value is about zero and a mean residue rotation at 233 nm, $[R']_{233}$, is about −2900 (Table IX-1). These two parameters, which are often used as a measure of the helix content of proteins, also indicate that protamine has a random-coil structure in aqueous solution (BRADBURY *et al.*, 1967).

Table IX-1. b_0 and R'_{233} values for protamines[a]

Solvent	b_0[b]			R'_{233}[b]
	Truttine	Clupeine	Salmine	Salmine
8 M Urea + H_2O	+ 5	+ 2	+ 16	−2850
0.25 M NaCl	+ 20	+ 19	+ 27	−2900
1 M NaCl	+ 3	0	+ 5	−2950
2-Chloroethanol	−370	−315	−375	−6600

[a] From BRADBURY *et al.*, 1967.
[b] b_0 values for the random and helical forms are thought to be 0 and −630°, respectively and R'_{233} values are −2500° and −13,500°, respectively.

Although protamine is insoluble in most of the helix solvents, it dissolves in 2-chloroethanol and dioxane. In these solvents it can form the helical structure (maximally ca. 40%), but it generally has random-coil structure in most other solvents [SUZUKI and ANDO, 1968 (1)].

Chapter X

Properties and Functions

A. Physical Properties

Nucleoprotamine dissolves in aqueous solution of NaCl (1—2 M) or ammonium sulfate to give a viscous solution. In such solutions it largely dissociates into DNA and protamine, but upon dilution (*e.g.*, to 0.14 M NaCl) they combine to form white fibrous precipitates [WATANABE and SUZUKI, 1951 (3)].

Table X-1. Binding of basic proteins to DNA[a]
Values of binding coefficient[b], $K_B(M^{-1})$, in 0.1 and in 0.95 M-NaCl

Proteins	Binding coefficients ($\times 10^2$)			
	In 0.1 M-NaCl Maximum binding		In 0.95 M-NaCl Minimum binding	
	Native DNA	Denatured DNA	Native DNA	Denatured DNA
Protamine	15.0	5.9	1.2	0.6
Histone IV	7.5	5.1	1.8	1.3
Histone I b	1.9	1.8	0.4	0.6
Poly-L-lysine	2.1	1.9	1.6	1.2

[a] From AKINRIMISI *et al.*, 1965.

[b] Binding coefficient ($K_B[M^{-1}]$) $= \dfrac{\text{Moles of protein amino nitrogen bound}}{\begin{pmatrix}\text{DNA concentration}\\ \text{in moles nucleotide/l}\end{pmatrix} \cdot \begin{pmatrix}\text{Moles of protein}\\ \text{amino nitrogen free}\end{pmatrix}}$

The dissociation into DNA and protamine can be demonstrated by the fact that protamines can be dialyzed from such solutions [AKINRIMISI *et al.*, 1965; WATANABE and SUZUKI, 1951 (1, 2)] and that DNA alone can be precipitated by addition of alcohol (HAMMARSTEN, 1924). Although the binding of protamine to DNA decreases sharply over the NaCl concentration range 0.3 to 0.7 M, the fact that protamine has an affinity for DNA even in 1.0 M NaCl indicates that interactions other than the main electrostatic one contribute to such binding. Protamine binds more strongly to native DNA than to denatured DNA. The affinity of protamine for native DNA is greater than that of histone and polylysine, as shown in Table X-1 (AKINRIMISI *et al.*, 1965).

The isoelectric point of most protamines is about twelve (MIYAKE, 1927). The values of specific rotation ($[\alpha]_D$) for clupeine and salmine are $-83.07°$ or $-84.0°$, and $-80.97°$, respectively (KOSSEL, 1898; WALDSCHMIDT-LEITZ *et al.*, 1926). The pK's of dissociable groups in clupeine and salmine are as follows; pK_1 (-COOH) 2.9—3.3; pK_2 ($-NH_2$ or $>$ NH) 7.4—8.2 or 8.9; and pK_3 (guanidino group) above twelve [HASHIMOTO, 1959 (2); RASMUSSEN and LINDERSTRØM-LANG, 1934].

The physico-chemical constants of clupeine and salmine are as follows: diffusion constant $D^0_{20,w} = 12 \times 10^{-7}$ cm²/sec, and intrinsic viscosity $[\eta]_v = 9$ (assuming partial specific volume $\bar{v} = 0.74$) in phosphate buffer, pH 7.7 ($\mu = 0.2$). Let us imagine protamine as an unhydrated ellipsoidal rod, the molecule being 13 Å wide and 90—100 Å long (ISO *et al.*, 1954). Iridine has a sedimentation constant $S^0_{20,w}$ of 0.74 S, a diffusion constant D^0_{20} of 13.0×10^{-7} cm²/sec, an intrinsic viscosity $[\eta]c$ of 8.16 and a partial specific volume $\bar{v}$ of 0.66 ml/g in 0.05 M triethanolamine hydrochloride buffer, pH 7.3. Assuming iridine behaves as a flexible rod, the diameter is 7.6 Å and the length is 104.5 Å (axial ratio 13.8) (GEHATIA and HASHIMOTO, 1963).

Ultraviolet-absorption spectra of protamines generally have no maximum in the region above 200 nm except for those containing aromatic amino acids, such as galline (DALY *et al.*, 1951; FISCHER and KREUZER, 1953; NAKANO *et al.*, 1968, 1969) and thynnine (BRETZEL, 1967). Clupeine has an absorption maximum at about 190 nm which is ascribable to the absorption of peptide bonds; the extinction coefficient of clupeine at this wavelength is about 7,000 (SUZUKI and ANDO, unpublished data). This value agrees with that obtained with synthetic polypeptides in random-coil structure (IMAHORI and TANAKA, 1959; ROSENHECK and DOTY, 1961). DNP-protamines show absorption spectra in the near ultraviolet region which are characteristic of the N-terminal amino acid [ANDO *et al.*, 1952, 1958 (2), HASHIMOTO, 1955; BRETZEL, 1967].

Infrared spectra (BUSWELL and GORE, 1942; KLOTS *et al.*, 1949; MIZUSHIMA *et al.*, 1950; BRADBURY *et al.*, 1967), optical rotatory dispersion (YANG and DOTY, 1957) and NMR spectra (BRADBURY *et al.*, 1967) of protamine have been measured.

B. Chemical Properties

Protamine has no tryptophan or sulfur-containing amino acids and most of it are also lacking in tyrosine and phenylalanine. Therefore reactions specific for such amino acids cannot be used for protamine. It is not antigenic, probably due to the absence of aromatic amino acids, and of any rigid structure. Since protamine is of low molecular weight and has no rigid structure, denaturation and coagulation do not occur upon heating.

As protamine has a high arginine content, it is easily digested by trypsin but not by pepsin [HASHIMOTO, 1959 (1); YAMASAKI, 1960]. It is also hydrolyzed to a certain extent by subtilisin or chymotrypsin (YAMASAKI, 1960). Thermolysin, a heat-stable protease from *B. stearothermoproteolyticus* [ANDO and SUZUKI, 1967; SUZUKI and ANDO, 1972 (2)], and a Neutral Protease from *B. subtilis* (ANDO and WATANABE, 1969), both of which hydrolyze protamine to a limited extent, are very useful in sequencing protamine when used in combination with trypsin (cf. Chap. VIII. C and F).

When clupeine or salmine is treated with concentrated sulfuric acid, N $\rightarrow$ O acyl rearrangement takes place at the N-terminal peptide bonds involving serine and threonine residues in a yield of about 70% (IWAI, 1959). In order to achieve selective cleavage at such peptide bonds, the rearranged protamine is first treated with acetic anhydride to block liberated amino groups and then with alkali to split the ester bonds (IWAI, 1961; IWAI and ANDO, 1967). This method has played an important role in the determination of chemical structures of protamine (ANDO *et al.*, 1962; ANDO and SUZUKI, 1966). The guanidino groups of arginine residues in protamine can be selectively modified by benzil (ITANO and GOTTLIEB, 1963) or diacetyl (YANKEELOV, Jr. *et al.*, 1966). The modified protamine is resistant to the action of trypsin (ITANO and GOTTLIEB, 1963).

Treatment of protamine with alkali results in racemization, hydrolysis of peptide bonds and decomposition of guanidino groups [KOSSEL and WEISS, 1909 (1, 2); 1910]. Therefore, on acid hydrolysis, alkali-treated protamine gives D,L-arginine and D,L-ornithine (DAKIN, 1912; DAKIN and DUDLEY, 1913). When hydrazine is used, arginine is converted into ornithine. This reaction can be used for the identification of the C-terminal amino acid of protamine (KAWANISHI *et al.*, 1964). During acid hydrolysis (6 N HCl, 110 °C, 24 h) about 1% of arginine is converted into ornithine (MURRAY *et al.*, 1965). The presence of traces of sulfate can sometimes lead to the formation of serine *O*-sulfate during the evaporation of acid hydrolysates (MURRAY and MILSTEIN, 1967).

In the SAKAGUCHI reaction using α-naphthol and hypobromite in alkaline solution (WEBER, 1930), 30—50% of the arginine residues in protamine are available (ROCHE and BLANC-JEAN, 1940). In the improved SAKAGUCHI reaction using 8-hydroxyquinoline (SAKAGUCHI, 1950, 1951), only about 20% of the arginine residues in clupeine, salmine and iridine react with the reagent [HASHIMOTO, 1959 (2); ISHII, 1960 (1)]. The relative color yields of the SAKAGUCHI reaction of arginine peptides of various lengths, $(Arg)_n$, are as follows; 100, 92, 64, 62, 60, 58, 56, 47, 47 and $< 10\%$ for $n = 1, 2, 3, 4, 5, 6, 7, 8, 9$ and ca. 90, respectively (KAWASHIMA, unpublished data; see also ITO *et al.*, 1970). However, in a newly modified form of the SAKAGUCHI reaction about 80% of the arginine residues in clupeine react (MESSINEO, 1966).

Treatment with hypobromite alone in alkaline solution leads to decomposition of guanidino groups with evolution of nitrogen gas. In this reaction, about 70% of the guanidino groups in salmine and clupeine can be modified [KIMURA, 1960 (1, 2)]. As described so far, arginine residues in protamines are less reactive than free arginine. This seems to be due to the steric hindrance of the guanidino groups of consecutive arginine residues [ISHII, 1960 (1)].

Nitration of most of the guanidino groups in protamine can be achieved by treatment with concentrated sulfuric acid and fuming nitric acid under cooling (KOSSEL and KENNAWAY, 1911; ROCHE and MOURQUE, 1948). The derivative thus formed gives nitroarginine on acid hydrolysis and ornithine on alkaline hydrolysis (KOSSEL and WEISS, 1913).

The C-terminal carboxyl group of protamine can be changed into its hydroxamate by reaction with hydroxylamine (TOBITA *et al.*, 1968).

Since protamine is strongly basic, it forms salt-like complexes with various inorganic and organic acids. The sulfate, hydrochloride, nitrate, phosphate, and acetate salts are soluble in water, but the picrate, flavianate, and phosphotungstate salts are almost insoluble in water. Protamines also form scarcely soluble salts with silver nitrate, copper salts, mercury salts, etc. (KOSSEL, 1929). When heated to 60 °C with the MIRSKY histone reagent (0.34 M mercuric sulfate in 1.88 M sulfuric acid) (DALY *et al.*, 1951), protamine and histone are not precipitated whereas many other proteins precipitate. If the mixture is cooled to 0 °C, the mercury salt of protamine is precipitated but histone remains in solution. This phenomenon can be utilized to distinguish protamine from histone.

In many cases, it is possible to change salt forms fairly easily by treatment with strongly basic ion-exchange resins (such as Amberlite IRA-400) (CALLANAN *et al.*,

1957) or weakly acidic ion-exchange resins (such as Amberlite XE-64) [Ishii, 1960 (2)]. It can also be done by the precipitation method (Callanan *et al.*, 1957).

Protamine also forms insoluble complexes with polysaccharides [Przylecki *et al.*, 1935 (1, 2); 1936], polysaccharide acids (Chargaff *et al.*, 1937, 1938), nucleic acids (Warburg and Christian, 1939), fatty acids (Jukes and Schmidt, 1935), proteins [Kossel, 1896 (2)], viruses (Chambers and Henle, 1941; Warren *et al.*, 1949) etc. Most of the precipitates, however, can be redissolved by the addition of salts. This property can therefore be utilized for removal of contaminants and for concentration of various materials during purification [Ross, 1954 (1, 2); Chamberlin and Berg, 1962; Philipps and West, 1964].

C. Biological and Physiological Properties

1. Biological Functions

Since histones were found to inhibit DNA-dependent RNA synthesis *in vitro* (Huang and Bonner, 1962), attention has been turned to the hypothesis that the basic nuclear proteins present in the cell nucleus in combination with DNA might be gene inhibitors or gene regulators (Stedman and Stedman, 1950, 1951). Experiments to test this hypothesis have been carried out, especially with histone, and are still in progress in several laboratories. As regards protamine, only a few results along these lines have been published, so that the biological function specific to protamine is still not clear at present.

Huang *et al.*, (1964) first reported that, unlike histone, protamine did not inhibit DNA-dependent RNA synthesis; however, all later results have shown that protamine strongly inhibits DNA-dependent RNA synthesis (Skalka *et al.*, 1966; Suzuki and Ando, 1969). Suzuki and Ando (1969) investigated the effect of the three homogeneous components of clupeine on DNA-dependent RNA synthesis using *Escherichia coli* RNA polymerase (Fig. X-1). The clupeine components inhibited RNA synthesis more strongly than the control homopolymers such as polylysine, polyornithine and polyarginine. There was also a slight difference in the extent of inhibition among the three components of clupeine, suggesting some differences in their mode of binding to DNA. The difference between polyarginine and clupeine in their effect on RNA synthesis indicates that RNA polymerase can distinguish the mode of their binding to DNA and that neutral amino acid residues as well as arginine residues may have some role in the binding of clupeine to DNA.

The biosynthetic development of protamine was first studied by analysis of amino-acid composition and terminal groups of proteins in the nuclei fraction from the testis of rainbowtrout *(Salmo irideus)* at different stages of spermatogenesis, as shown in Table X-2. These studies showed that in the immature stage DNA is bound to basic proteins of the histone type, which are gradually replaced as maturation proceeds by basic proteins of the protamine type, so that in the fully matured sperm heads DNA is bound only to protamine [Ando and Hashimoto, 1958 (1 2, 3); Felix *et al.*, 1958; Felix, 1960].

Ingles *et al.* (1966) induced spermatogenesis in immature steelhead trout by injections of salmon pituitary extracts. The injections cause a rapid increase in the weight of testes. The other biochemical changes in these testes are the same as those of the adult fish during the spawning migration. This procedure makes it possible to study the biosynthesis of protamine at any time of the year. Dixon and his co-workers have also established a system for cell-free synthesizing nuclear basic proteins (LING *et al.*, 1969). By the use of these systems they have established the following facts: at a late stage of spermatogenesis protamine appears in the nuclei and this newly synthesized protamine progressively replaces histone in combination with DNA (INGLES *et al.*, 1966; LING *et al.*, 1969). When protamine synthesis begins, histone synthesis declines and eventually ceases. Lastly, in the trout lysine-rich (FI) histone is replaced by protamine (MARUSHIGE and DIXON, 1969), but in an *in vitro* experiment

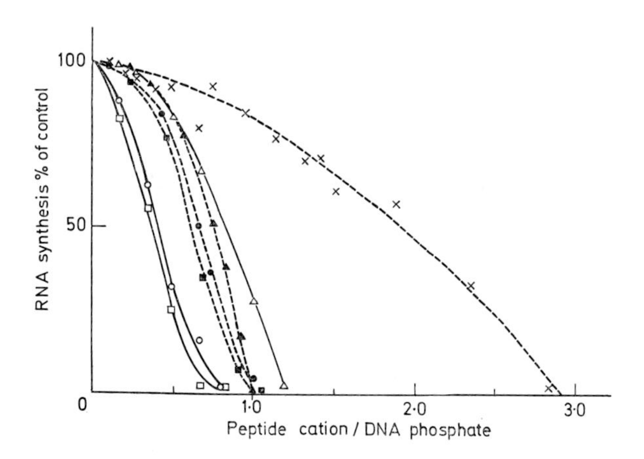

Fig. X-1. Inhibition of RNA synthesis by clupeine components and basic homopolypeptides. — The RNA synthesizing system contains a total volume of 0.5 ml aqueous solution. Reagents are added in the following order: 20 µmoles of Tris-HCl, pH 8.1; 6 µmoles of β-mercaptoethanol; 2 µmoles of $MgCl_2$; 0.5 µmole of $MnCl_2$; 190 mµmoles of DNA, clupeine or basic homopolypeptides as indicated in the figure; 100 mµmoles each of ^{14}C-UTP, cold ATP, GTP and CTP; and about 20 µg of RNA polymerase from *E. coli* B. After 10 min incubation at 37 °C, acid-insoluble material is collected and the radioactivity is measured. —— ○ —— Clupeine YI, —— □ —— YII, —— △ —— Z, --- ● --- Poly-L-Arg (degree of polymerization, DP ≑ 90), --- ■ --- Poly-L-Lys (DP ≑ 100), --- ▲ --- Poly-L-Orn (DP = 90). Only in the case when poly-L-Arg is added after the addition of 4 XTP (*i.e.*, DNA-poly-L-Arg complex is formed in the presence of 4 XTP), is the inhibitory effect markedly changed, as shown by the curve --- × --- (reproduced by combining 2 figures from SUZUKI and ANDO, 1969)

using chicken erythrocyte chromatin added to protamine, it was replaced by arginine-rich histone (FIV) (EVANS *et al.*, 1970).

Protamine is synthesized in cytoplasm on diribosomes (disomes) which are present in high concentration during the period of rapid protamine synthesis.. These disomes incorporate arginine rapidly but lysine very slowly, which suggests that they are principally engaged in protamine synthesis (LING *et al.*, 1969; LING and DIXON, 1970). Inhibition experiments indicate that protamine is synthesized by the classical pathway involving *m*-RNA, *t*-RNA, and ribosomes (INGLES *et al.*, 1966). However, since actinomycin D has an inhibitory effect on the biosynthesis only during the early stage of protamine synthesis, *m*-RNA for protamine must be synthesized before actual protamine synthesis takes place and is presumed to be metabolically stable (LING and DIXON, 1970).

The newly synthesized protamine is phosphorylated by adenosine triphosphate in cytoplasm catalyzed by protamine kinase (LANGAN and SMITH, 1967; INGLES and DIXON, 1967; MARUSHIGE *et al.*, 1969; JERGIL and DIXON, 1970). The phosphorylation occurs at the hydroxyl groups of serine residues (TREVITHICK *et al.*, 1967; INGLES and DIXON, 1967). Protein kinase has been found in various mammalian tissues

Table X-2. Changes in the amino-acid composition[a], and N- and C-terminals of basic proteins in isolated nuclei from testes of 4—5 year old rainbow trout during the formation of spermatozoa [ANDO and HASHIMOTO, 1958 (1, 2)]

	Sept.	Oct.	Nov.	Dec.	Jan.-Mar.
Total N (%)	16.8	19.5	20.5	22.6	23.0
Arg	+ +	+ + + +	+ + + + +	+ + + + +	+ + + + +
Ala	+ + +	+ + +	+ + +	+	+
Leus	+ + +	+ + +	+ +	+	+
Thr	+	+ +	±	—	—
Lys	+ + + +	+ + +	+ +	±	—
Glu	+ +	+ +	+	—	—
Asp	+	+ +	+	—	—
His	+	+	±	—	—
Tyr	+	+	—	—	—
Phe	+	+	—	—	—
(Cys)[b]	+	+	+	—	—
N-Terminal[c,d]	Ala, (Pro)[g]	Ala, Pro	Pro, (Ala)[g]	Pro	Pro
(ε-DNP-Lys)[c]	(present)	(present)	(present)	(trace)	(absent)
C-Terminal[e,f]			Arg		Arg

[a] In addition to these amino acids, Gly, Ser, Val and Pro were always present (+ + + or + +). Trp was absent. + or — shows relative color yield by ninhydrin reaction on a two-dimensional paper chromatogram per hydrolyzate of a unit weight of protein.
[b] The spot of cystine is somewhat uncertain.
[c] By the DNP-method.
[d] By the PTC-method.
[e] By digestion with carboxypeptidase B fraction.
[f] By the hydrazinolysis method.
[g] Parenthesized amino acid was detected in slight quantity.

[LANGAN and SMITH, 1967; WALSH *et al.*, 1968; MIYAMOTO *et al.*, 1969 (1, 2); KUO and GREENGARD, 1969 (1)], in invertebrate phyla [KUO and GREENGARD, 1969 (1)], and in bacteria [KUO and GREENGARD, 1969 (2)]. The phosphorylation of histone and protamine by the kinase is known to be dependent on adenosine 3',5'-cyclic phosphate which markedly increases the rate of conversion [LANGAN, 1968; WALSH *et al.*, 1968; KUO and GREENGARD, 1969 (1, 2); MIYAMOTO *et al.*, 1969 (1, 2); MARUSHIGE *et al.*, 1969; JERGIL and DIXON, 1970]. Modification of protamine by phosphorylation of its serine residues would considerably reduce the net charge of protamine and hence presumably alter its interaction with DNA. The phosphorylated protamine is transferred into the cell nucleus and then into the chromatin, where it replaces the histone and binds to DNA without appreciable dephosphorylation (MARUSHIGE *et al.*, 1969). A phosphatase specific to phosphorylated protamine and histone has been found in rat liver (MEISLER and LANGAN, 1969). An enzyme of this kind must be present in mature testes to dephosphorylate protamine in the DNA-phosphorylated protamine complex in order to give the DNA-protamine found in mature sperm heads.

The synthesis of protamine in steelhead trout starts from methionine, *i.e.* methionine acts as chain initiator in protamine synthesis, as it does in protein synthesis in bacterial systems (WIGLE and DIXON, 1970). Methionine is incorporated into every component of iridine, constituting some 70% of the total amount of N-terminal amino acids. After tryptic digestion of newly synthesized protamine, a peptide Met-Pro-Arg has been found. The proline residues seems to be the N-terminus of mature protamine. The methionine residue is removed after chain completion by an enzyme found in the extract of testis cells.

Mainly by means of cytochemical methods, a similar process of replacement of histone-type proteins by protamine-type proteins has been observed in sperm cells at the later stages of spermatogenesis in many species of organisms, *e.g.*, *Helix aspersa* [BLOCH and HEW, 1960 (1,2)], *Drosophila melanogaster* [DAS *et al.*, 1964 (1,2,3)], mouse (MONESI, 1964, 1965), rat (VAUGHN, 1966) and grasshopper (BLOCH and BRACK, 1964). Generally, once the proteins in the sperm nucleus have been completely replaced by those of protamine type, the metabolic activity in the sperm ceases.

The results described so far indicate that the genetic information in DNA is completely shut off when protamine binds to DNA. Therefore the role of protamine might be to protect DNA from degradation and to keep its structure compact for transfer into egg cells (FELIX *et al.*, 1956; FELIX, 1960). Why, then, must several components be present in a protamine? Do they all play only a passive role, or do they have some other important role after fertilization? These interesting questions remain to be answered.

In an attempt to solve these questions, studies of DNA-protamine complexes have been made from the point of view of thermal denaturation (HUANG *et al.*, 1964; RAUKAS, 1965, 1966; INOUE and ANDO, 1966; OLINS *et al.*, 1968; see also LENG and FELSENFELD, 1966). Fig. X-2 shows the thermal denaturation curves of DNA-clupeine complexes (INOUE and ANDO, 1966). HUANG *et al.* (1964) first reported that the denaturation curve of a DNA-salmine (commercial) complex was identical to that of DNA. However, when protamine is added to a solution of DNA in an amount less than that of the DNA, two Tm values can generally be observed. The lower Tm value is thought to reflect the denaturation of free DNA, and the higher one that of a DNA-protamine complex. Thus protamine increases the thermal stability of DNA when it binds to it. The amount of DNA denatured at the higher Tm value is in proportion to the amount of protamine added (insets, Fig. X-2 a and b). If curve is extrapolated to the point where the Tm of free DNA disappears (100% bound DNA), which may represent the state of native nucleoprotamine, the molar ratio of arginine in clupeine to phosphate in DNA is 0.8 or 1.0, depending on the method of inducing complex formation.

Studies on the interaction between DNA and protamine or basic homopolypeptides were carried out in parallel with the Tm measurements, and protamine was shown to bind to DNA stoichiometrically and cooperatively (OLINS *et al.*, 1967, 1968; AKINRIMISI *et al.*, 1965; SUZUKI and ANDO, 1969; KAWASHIMA *et al.*, 1969). Polylysine and polyarginine were also shown to bind fairly selectively to the AT-rich and GC-rich regions in DNA, respectively (LENG and FELSENFELD, 1966). However, it has not yet been established whether any of the components of protamine interacts with a specific site in DNA (OHBA, 1966).

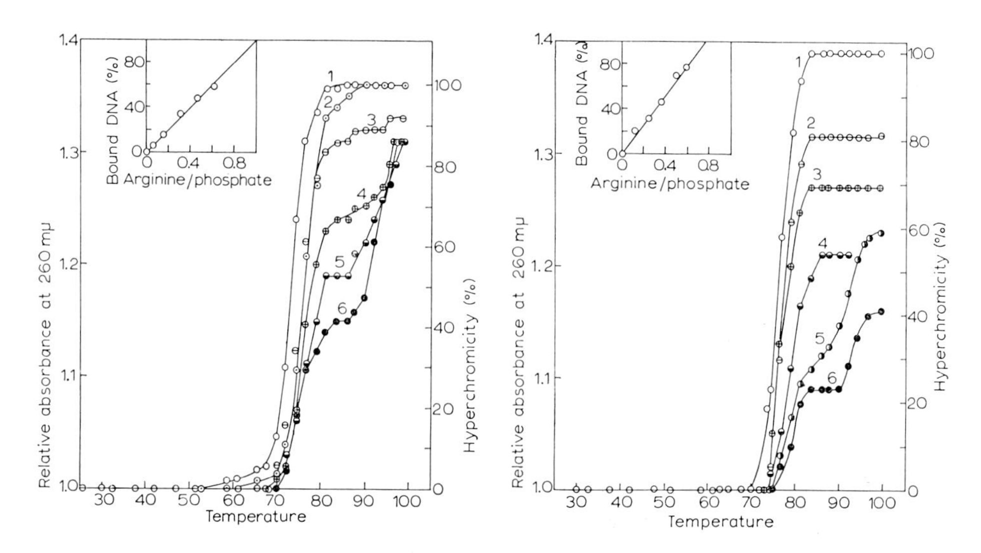

Fig. X-2 a (left). Absorbance-temperature profile of DNA-clupeine formed by "rapid binding". DNA concentration, $7.6 \cdot 10^{-5}$ M (phosphate); Curve 1, DNA alone; the arginine/phosphate ratios are for Curve 2, 0.06; Curve 3, 0.16; Curve 4, 0.31; Curve 5, 0.47; Curve 6, 0.62

Fig. X-2 b (right). Absorbance-temperature profile of DNA-clupeine formed by "slow binding". DNA concentration, $5.5 \cdot 10^{-5}$ M (phosphate); Curve 1, DNA alone; the arginine/phosphate ratios are for Curve 2, 0.12; Curve 3, 0.24; Curve 4, 0.36; Curve 5, 0.50; Curve 6, 0.59

Reproduced from INOUE and ANDO (1966). "Rapid binding" means simply to mix the solutions of clupeine and DNA in 0.03 M NaCl-0.003 M sodium citrate, pH 7.0, by titrating the latter solution with the former. "Slow binding" means to make a nucleoclupeine complex by successive dialysis of a mixture of solutions of both nucleic acid and clupeine in 2 M NaCl against 2 M, 0.4 M, 0.15 M NaCl and finally against 0.03 M NaCl containing 0.003 M sodium citrate, pH 7.0

Caution is necessary in comparing the experimental results obtained with a reconstituted system and with native nucleoprotamines, since there is some evidence that reconstituted DNA-histone is different in nature from the native form (JOHNS and BUTLER, 1964).

In addition to DNA and protamine, native chromatin contains so-called residual proteins [MIRSKY, 1947; MIRSKY and RIS, 1947 (1, 2); STEDMAN and STEDMAN 1947, (1, 2); see also Chap. II]. Although little is known about their nature, they may play some essential role in the structure and function of the nucleoprotamine. This question awaits further studies.

2. Physiological Properties

It was reported that plasmolyzed sperm nuclei from a species of trout were still capable of fertilizing eggs in 45% yield, and that the young fish so bred were scarcely distinguishable from normally bred ones [FELIX *et al.*, 1952 (2); SCHNEIDER, 1954].

Protamines have also been reported to show a variety of biological and physiological activities. The following effects have been observed, for example: inhibitory effects on respiration, digestion, blood pressure and cardiac activity in various animals (HEIM, 1943; THOMPSON, 1900; JAQUES, 1949); influences on blood coagulation (THOMPSON, 1900; WALDSCHMIDT-LEITZ *et al.*, 1929); prolonged lowering of blood sugar with few secondary reactions when used together with zinc-insulin (HAGEDORN *et al.*, 1936; MARK and BISKIND, 1940), but elevation of sugar content when used alone in animals (SCHENCK, 1932); also some effects before and after fertilization [FELIX *et al.*, 1936, 1952 (2); FELIX, 1952; SCHNEIDER, 1954]. More recently, protamine has been reported to stimulate the incorporation into cells of RNA (AMOS and KEARNS, 1963) and DNA (MENDECKI and WILCZOK, 1963; WILCZOK and MENDECKI, 1963); to increase the plaque-forming capacity and infectivity of polio virus (SMULL *et al.*, 1961; SMULL and LUDWIG, 1962; CONNOLLY, 1966) but to inhibit the adsorption of viruses and phages (COLTER *et al.*, 1964; COLTER and CAMPBELL, 1965; ZICKLER, 1967), and to be an antitumor agent (GARVIE, 1965; BRADSHAW and ROSSER, Jr., 1969). Further, it is known to inhibit cell growth (LISNELL and MELLGREN, 1963), to transform cells (LALLIER, 1963), and to inhibit egg cleavage (CORMACK, 1966) and mitosis (BELL and HESLOT, 1969).

Protamine is also known to bind to the cell membrane, increasing its permeability (LEITCH and TOBIAS, 1964; LARSEN, 1967) and to bind to enzymes, modifying their activity. Thus, protamines are known to inactivate the following enzymes: D,L-lactic cytochrome *c* reductase (NYGAARD, 1962), phospholipase D (DAWSON and HEMINGTON, 1967) and lipase (PAYZA *et al.*, 1967). Activation by protamines of the following enzymes is reported: Na^+, K^+-dependent ATPase (YOSHIDA *et al.*, 1965) and phosphoenolpyruvate carboxylase (SANWAL *et al.*, 1966). These effects, however, are thought to be due to the strong basicity of protamines and therefore should be considered separately from their biological functions.

Acknowledgement

The authors are greatly indebted to our colleagues, Dr. KENJI TAKAHASHI, Dr. SEIICHI KAWASHIMA, Mrs. KAZUKO KAWASHIMA, Mrs. YOKO SHIGA and Miss MIDORI IWATANI, for their helpful collaboration in preparing the manuscript.

References

Akabori, S., Ohno, K., Narita, K.: On the hydrazinolysis of proteins and peptides: A method for the characterization of carboxyl-terminal amino acids in proteins. Bull. Chem. Soc. Japan **25**, 214—218 (1952).

Akabori, S., Ohno, K., Ikenaka, T., Nagata, A., Haruna, I.: Improved methods for the characterization and quantitative estimation of carboxyl-terminal amino acids in proteins. Proc. Jap. Acad. **29**, 561—564 (1953).

Akinrimisi, E. O., Bonner, J., Ts'o, P. O. P.: Binding of basic proteins to DNA. J. molec. Biol. **11**, 128—136 (1965).

Alexander, P.: The combination of protamine with desoxyribonucleic acid. Biochim. biophys. Acta (Amst.) **10**, 595—599 (1953).

Allfrey, V. G.: Control mechanisms in ribonucleic acid synthesis. Cancer Res. **26**, 2026—2040 (1966).

Amos, H., Kearns, K. E.: Influence of bacterial ribonucleic acid on animal cells in culture. II. Protamine enhancement of RNA uptake. Exp. Cell Res. **32**, 14—25 (1963).

Ando, T.: Studies on the chemical structure of protamines. II. (in Japanese). Kagaku-to-Kogyo (Chemistry and Industry) **17**, 466—479 (1964).

Ando, T., Hashimoto, C.: Chemical studies on proteins. III. The carboxyl-terminal group of clupeine (in Japanese). Rep. Radiat. Chem. Res. Inst., Univ. Tokyo No. 5, 56—59 (1950).

Ando, T., Hashimoto, C.: (1) Studies on protamines. IV. The basic proteins of the testis cell nuclei of adult and immature rainbow-trouts in breeding season. J. Biochem. (Tokyo) **45**, 453—460 (1958).

Ando, T., Hashimoto, C.: (2) Studies on protamines. V. Changes of the proteins in the cell nuclei of the testis during the formation of spermatozoa of the rainbow-trout *(Salmo irideus)*. J. Biochem. (Tokyo) **45**, 529—540 (1958).

Ando, T., Hashimoto, C.: (3) Changes of the proteins in the cell nuclei of the testis during the formation of spermatozoa of the rainbow-trout *(Salmo irideus)*. Proc. Intern. Symp. Enz. Chem., Tokyo and Kyoto 1957, 380—385 (1958).

Ando, T., Sawada, F.: On the heterogeneity of protamines (clupeine and iridine) obtained from spermatozoa of each single fish. J. Biochem. (Tokyo) **46**, 517—519 (1959).

Ando, T., Sawada, F.: Studies on protamines. VIII. Chromatographic studies on the heterogeneity of iridine. J. Biochem. (Tokyo) **48**, 886—898 (1960).

Ando, T., Sawada, F.: Studies on protamines. IX. Fractionation of clupeine. J. Biochem. (Tokyo) **49**, 252—259 (1961).

Ando, T., Sawada, F.: Studies on protamines. X. Heterogeneity of clupeine from single herring. J. Biochem. (Tokyo) **51**, 416—421 (1962).

Ando, T., Suzuki, Ko.: The amino acid sequence of the second component of clupeine. Biochim. biophys. Acta (Amst.) **121**, 427—429 (1966).

Ando, T., Suzuki, Ko.: The amino acid sequence of the third component of clupeine. Biochim. biophys. Acta (Amst.) **140**, 375—377 (1967).

Ando, T., Watanabe, S.: A new method for fractionation of protamines and the amino acid sequences of one component of salmine and three components of iridine. Int. J. Protein Res. **1**, 221—224 (1969).

Ando, T., Ishii, S., Hashimoto, C., Yamasaki, M., Iwai, K.: On the constituent amino acids, N-terminal residues and the molecular weights of protamines. Bull. Chem. Soc. Japan **25**, 132 (1952).

Ando, T., Iwai, K., Yamasaki, M., Hashimoto, C., Kimura, M., Ishii, S., Tamura, T.: Further notes on protamines. Bull. Chem. Soc. Japan **26**, 406—407 (1953).

ANDO, T., IWAI, K., ISHII, S., YAMASAKI, M., KIMURA, M., TOBITA, T., SATO, M., ABUKUMAGAWA, E.: Chemical structure of protamines — Studies by partial hydrolysis (in Japanese). Proc. 7th Symposium on Protein Structure, Tokyo, Nov. 4, pp. 11—12, 1955.

ANDO, T., ISHII, S., YAMASAKI, M., TOBITA, T., SAWADA, F., IWAI, K., FUJIOKA, H., KIMURA, M., SATO, M., ABUKUMAGAWA, E., KAWANISHI, Y.: (1) Chemical structure of clupeine and salmine. Proc. Symposium on Protein Structure, Tokyo, Oct. 28, pp. 1—17 (1957).

ANDO, T., ABUKUMAGAWA, E., NAGAI, Y., YAMASAKI, M.: (2) On the N-terminal residues and sequence of clupeine from *Clupea pallasii*: The occurrence of proline in addition to alanine as the N-terminus. J. Biochem. (Tokyo) **44**, 191—194 (1957).

ANDO, T., ISHII, S., YAMASAKI, M., IWAI, K., HASHIMOTO, C., SAWADA, F.: (3) Studies on protamines. I. Amino acid composition and homogeneity of clupeine, salmine and iridine. J. Biochem. (Tokyo) **44**, 275—288 (1957).

ANDO, T., NAGAI, Y., FUJIOKA, H.: The action of leucine aminopeptidase on protamines. J. Biochem. (Tokyo) **44**, 779—781 (1957).

ANDO, T., TOBITA, T., YAMASAKI, M.: (1) Effects of carboxypeptidases (Anson's and Basic) on protamines. J. Biochem. (Tokyo) **45**, 285—290 (1958).

ANDO, T., YAMASAKI, M., ABUKUMWGAWA, E., ISHII, S., NAGAI, Y.: (2) Studies on protamines. III. N-Terminal residues of salmine and clupeine. J. Biochem. (Tokyo) **45**, 429—451 (1958).

ANDO, T., ISHII, S., YAMASAKI, M.: Peptides obtained by tryptic digestion of clupeine. Biochim. biophys. Acta (Amst.) **34**, 600—601 (1959).

ANDO, T., IWAI, K., ISHII, S., AZEGAMI, M., NAKAHARA, C.: The chemical structure of one component of clupeine. Biochim. biophys. Acta (Amst.) **56**, 628—630 (1962).

ANDO, T., SUZUKI, KO., WATANABE, S., INOUE, S.: Studies on protamines. Abstr. I, Symposium I, 7th Intern. Congr. Biochem. 1—1, Tokyo: The Science Council of Japan 1967.

AZEGAMI, M.: The partial chemical structure of the clupeine Z component. Dissert. for a degree of M. Sci., Univ. Tokyo, March 1961.

AZEGAMI, M., ISHII, S., ANDO, T.: Studies on protamines. XIV. Partial chemical structure of one component of clupeine, clupeine Z. J. Biochem. (Tokyo) **67**, 525—534 (1970).

BELL, S., HESLOT, H.: Effect of protamine sulfate in cells of *Vicia faba* roots. Exp. Cell Res. **54**, 77—82 (1969).

BLACK, J. A., DIXON, G. H.: Evolution of protamine: A further example of partial gene duplication. Nature (Lond.) **216**, 152—154 (1967).

BLOCH, D. P., BRACK, S. D.: Evidence for the cytoplasmic synthesis of nuclear histone during spermiogenesis in grasshopper *Chortophaga viridifasciata* (de Geer). J. Cell Biol. **22**, 327—340 (1964).

BLOCH, D. P., HEW, H. Y. C.: (1) Schedule of spermatogenesis in the pulmonate snail, *Helix aspersa*, with special reference to histone transition. J. biophys. biochem. Cytol. **7**, 515—532 (1960).

BLOCH, D. P., HEW, H. Y. C.: (2) Changes in nuclear histones during fertilization, and early embryonic development in the pulmonate snail, *Helix aspersa*. J. biophys. biochem. Cytol. **8**, 69—81 (1960).

BLOCK, R. J., BOLLING, D., GERSHON, H., SOBER, H. A.: Preparation and amino acid composition of salmine and clupeine. Proc. Soc. exp. Biol. (N. Y.) **70**, 494—496 (1949).

BONNER, J., HUANG, R.-C. C.: Role of histone in chromosomal RNA synthesis. In: The nucleohistones (BONNER, J., Ts'o, P. O. P., Eds.), pp. 251—261. San Francisco: Holden-Day Inc. 1964.

BONNER, J., DAHMUS, M. E., FAMBROUGH, D., HUANG, R.-C. C., MARUSHIGE, K., TUAN, D. Y. H.: (1) The biology of isolated chromatin. Chromosomes, biologically active in the test tube, provide a powerful tool for the study of gene action. Science **159**, 47—56 (1968).

BONNER, J., CHALKLEY, C. R., DAHMUS, M. E., FAMBROUGH, D., FUJIMURA, F., HUANG, R.-C. C., HUBERMAN, J., JENSEN, R., MARUSHIGE, K., OHLENBUSCH, H., OLIVERA, B., WIDHOLM, J.: (2) Isolation and characterization of chromosomal nucleoproteins. In: Methods in enzymology (GROSSMAN, L., MOLDAVE, K., Eds.), Vol. XII, Part B, pp. 3—65. New York: Academic Press 1968.

BRADBURY, E. M., PRICE, W. C., WILKINSON, G. R.: (1) Polarized infrared studies on nucleoproteins. I. Nucleoprotamine. J. molec. Biol. **4**, 39—49 (1962).

BRADBURY, E. M., PRICE, W. C., WILKINSON, G. R., ZUBAY, G.: (2) Polarized infrared studies of nucleoproteins. II. Nucleohistone. J. molec. Biol. **4**, 50—60 (1962).

BRADBURY, E. M., CRANE-ROBINSON, C., GOLDMAN, H., RATTLE, H. W. E., STEPHENS, R. M.: Spectroscopic studies of the conformations of histones and protamine. J. molec. Biol. **29**, 507—523 (1967).

BRADSHAW, L. J., ROSSER, L. H., Jr.: Inhibition of Ehrlich ascites tumor growth by nucleoprotein. I. *In vivo* characterization and presumptive chemical identification. J. nat. Cancer Inst. **43**, 521—526 (1969).

BRETZEL, G.: Über Thynnin: Fraktionierung und Isolierung einer einheitlichen Komponente. VIII. Mitteilung über die Struktur der Protamine in der Untersuchungsreihe von E. WALDSCHMIDT-LEITZ u. Mitarb. Z. physiol. Chem. **348**, 419—425 (1967).

BRETZEL, G.: Über Thynnin, das Protamin des Thunfisches. IX. Mitteilung über die Struktur der Protamine in der Untersuchungsreihe von E. WALDSCHMIDT-LEITZ u. Mitarb. Z. physiol. Chem. **352**, 1025—1033 (1971).

BRETZEL, G.: (1) Über Thynnin, das Protamin des Thunfisches. Isolierung und Charakterisierung der aus Thynnin Y2 nach Einwirkung von Thermolysin erhaltenen Peptide. X. Mitteilung über die Struktur der Protamine in der Untersuchungsreihe von E. WALDSCHMIDT-LEITZ u. Mitarb. Z. physiol. Chem. **353**, 209—216 (1972).

BRETZEL, G.: (2) Über Thynnin, das Protamin des Thunfisches. Die vollständige Aminosäuresequenz von Thynnin Y2. XI. Mitteilung über die Struktur der Protamine in der Untersuchungsreihe von E. WALDSCHMIDT-LEITZ u. Mitarb. Z. physiol. Chem. **353**, 933—943 (1972).

BRETZEL, G.: (3) Über Thynnin, das Protamin des Thunfisches. Die Sequenz der Komponente Y1. XII. Mitteilung über die Struktur der Protamine in der Untersuchungsreihe von E. WALDSCHMIDT-LEITZ u. Mitarb. Z. physiol. Chem. **353**, 1362—1364 (1972).

BRIL-PETERSEN, E., WESTENBRINK, H. G. K.: A structural basic protein as a counterpart of deoxyribonucleic acid in mammalian spermatozoa. Biochim. biophys. Acta (Amst.) **76**, 152—154 (1963).

BUSCH, H.: Histones and other nuclear proteins, 121—149. New York: Acad. Press Inc. 1965.

BUSWELL, A. M., GORE, R. C.: Quantitative spectroscopic analysis of proteins. J. phys. Chem. **46**, 575—581 (1942).

CALLANAN, M. J., CARROLL, W. R., MITCHELL, E. R.: Physical and chemical properties of protamine from the sperm of salmon *(Oncorhynchus tschawytscha)*. I. Preparation and characterization. J. biol. Chem. **229**, 279—287 (1957).

CARTER, R. O.: The physical-chemical investigation of certain nucleoproteins. III. Molecular kinetic studies with calf thymus nucleohistone. J. Amer. chem. Soc. **63**, 1960—1964 (1941).

CARTER, R. O., HALL, J. L.: The physical chemical investigation of certain nucleoproteins. I. Preparation and general properties. J. Amer. chem. Soc. **62**, 1194—1196 (1940).

CHAMBERLIN, M., BERG, P.: Deoxyribonucleic acid-directed synthesis of ribonucliec acid by an enzyme from *Escherichia coli*. Proc. Nat. Acad. Sci. **48**, 81—94 (1962).

CHAMBERS, L. A., HENLE, W.: Precipitation of active influenza A virus from extra-embryonic fluids by protamine. Proc. Soc. exp. Biol. (N. Y.) **48**, 481—483 (1941).

CHANG, W. J.: Studies on the chemical structure of clupeine from North European herring. Dissert. for a degree of M. Sci., Univ. Tokyo, Feb. 1969.

CHARGAFF, E.: Studies on the chemistry of blood coagulation. VII. Protamines and blood clotting. J. biol. Chem. **125**, 671—676 (1938).

CHARGAFF, E., OLSON, K. B.: Studies on the chemistry of blood coagulation. VI. Studies on the action of heparin and other anticoagulants. The influence of protamine on the anticoagulant effect *in vivo*. J. biol. Chem. **122**, 153—167 (1937/38).

CHINARD, F. P.: Photometric estimation of proline and ornithine. J. biol. Chem. 91—95 (1952).

COELINGH, J. P.: Het histon van stierespermatozoën — isolatie, zuivering en karakterisering. Dissertation, The State University, Utrecht (The Netherlands), Oct. 1971.

COELINGH, J. P., ROZIJN, T. H., MONFOORT, C. H.: Isolation and partial characterization of a basic protein from bovine sperm heads. Biochim. biophys. Acta (Amst.) **188**, 353—356 (1969).

COELINGH, J. P., MONFOORT, C. H., ROZIJN, T. H., LEUVEN, J. A. G., SCHIPHOF, R., STEYN-PARVÉ, E. P., BRAUNITZER, G., SCHRANK, B., RUHFUS, A.: The complete amino-acid sequence of the basic nuclear protein of bull spermatozoa. Biochim. biophys. Acta (Amst.) **285**, 1—14 (1972).

COLTER, J. S., CAMPBELL, J. B.: Effect of polyanions and polycations on Mengo virus-L cell interaction. Ann. N. Y. Acad. Sci. **130**, 383—389 (1965).

COLTER, J. S., DAVIES, M. A., CAMPBELL, J. B.: Three variants of Mengo encephalomyelitis virus. II. Inhibition of interaction with L cells by an agar inhibitor and by protamine. Virology **24**, 578—585 (1964).

CONNOLLY, J. H.: Effect of histones and protamine on the infectivity of Semliki Forest virus and ribonucleic acid. Nature (Lond.) **212**, 858 (1966).

CORMACK, D. H.: Site of action of ribonuclease during its inhibition of egg cleavage. Nature (Lond.) **209**, 1364—1365 (1966).

DAIMLER, B. H.: (1) Untersuchungen über Desoxyribonucleinsäure und Clupein in der Ultrazentrifuge unter Verwendung einer neuen Universalregistriermethode. I. Teil: Die Universalregistriermethode. Kolloid-Z. **127**, 8—19 (1952).

DAIMLER, B. H.: (2) Untersuchungen über Desoxyribonucleinsäure und Clupein in der Ultrazentrifuge unter Verwendung einer neuen Universalregistriermethode. II. Teil: Die Ultrazentrifugenanlage, Untersuchungen über Desoxyribonucleinsäure und Clupein. Kolloid-Z. **127**, 97—110 (1952).

DAKIN, H. D.: The racemization of proteins and their derivatives resulting from tautomeric change. Part I. J. biol. Chem. **13**, 357—362 (1912).

DAKIN, H. D., DUDLEY, H. W.: The racemization of proteins and their derivatives resulting from tautomeric change. Part II. The racemization of casein. J. biol. Chem. **15**, 263—269 (1913).

D'ALCONTRES, G. S.: Protamines. G. Biochim. **4**, 128 (1955).

DALLAM, R. D., THOMAS, L. E.: Chemical studies on mammalian sperm. Biochim. biophys. Acta (Amst.) **11**, 79—89 (1953).

DALY, M. M., MIRSKY, A. E., RIS, H.: The amino acid composition and some properties of histones. J. gen. Physiol. **34**, 439—462 (1951).

DAS, C. C., KAUFMANN, B. P., GAY, H.: (1) Autoradiographic evidence of synthesis of an arginine-rich histone during spermiogenesis in *Drosophila melanogaster*. Nature (Lond.) **204**, 1008—1009 (1964).

DAS, C. C., KAUFMANN, B. P., GAY, H.: (2) Histone-protein transition in *Drosophila melanogaster*. I. Changes during spermatogenesis. Exp. Cell Res. **35**, 507—514 (1964).

DAS, C. C., KAUFMANN, B. P., GAY, H.: (3) Histone-protein transition in *Drosophila melanogaster*. II. Changes during early embryonic development. J. Cell Biol. **23**, 423—430 (1964).

DAS, N. K., SIEGEL, E. P., ALFERT, M.: Synthetic activities during spermatogenesis in the locust. J. Cell Biol. **25**, 387—395 (1965).

DAWSON, R. M. C., HEMINGTON, N.: Some properties of purified phospholipase D and especially the effect of amphipathic substances. Biochem. J. **102**, 76—86 (1967).

DIRR, K., FELIX, K.: (1) Über Clupein. III. Mitteilung. Z. physiol. Chem. **205**, 83—92 (1932).

DIRR, K., FELIX, K.: (2) Über Clupein. IV. Mitteilung. Z. physiol. Chem. **209**, 5—11 (1932).

ENDO, S.: Studies on protease produced by thermophilic bacteria (in Japanese). Hakko-Kogaku Zasshi (J. Ferment. Technol.) **40**, 346—353 (1962).

EVANS, K., KONIGSBERG, P., COLE, R. D.: Displacement of histones from deoxyribonucleoprotein by protamine. Arch. Biochem. Biophys. **141**, 389—392 (1970).

FELIX, K.: Über Clupein. Ber. ges. Physiol. **61**, 349—350 (1931).

FELIX, K.: Zur Chemie des Zellkerns. Experientia (Basel) **8**, 312—317 (1952).

FELIX, K.: Protamines and nucleoprotamines. In: The chemical structure of proteins (WOLSTENHOLME, G. E. W., CAMERON, M. P., Eds.) (Ciba Foundation Symposium), pp. 151—164. London: J. and A. Churchill 1953.

FELIX, K.: Protamines. Advanc. Protein Chem. **15**, 1—56 (1960).

FELIX, K., DIRR, K.: Über Clupein. 1. Mitteilung. Z. physiol. Chem. **184**, 111—131 (1929).

FELIX, K., HASHIMOTO, C.: Erneute Untersuchungen über die chemische Struktur von Protaminen. Z. physiol. Chem. **330**, 205—211 (1963).
FELIX, K., KREKELS, A.: Nucleoprotamin. V. Mitteilung. Z. physiol. Chem. **293**, 284—286 (1953).
FELIX, K., LANG, A.: Über Clupein. II. Mitteilung. Z. physiol. Chem. **193**, 1—14 (1930).
FELIX, K., MAGER, A.: Über Clupein. VIII. Mitteilung Z. physiol. Chem. **249**, 111—123 (1937).
FELIX, K., SCHNEIDER, H.: Zur Spezifität der Arginase. Z. physiol. Chem. **255**, 132—144 (1938).
FELIX, K., MÜLLER, H., DIRR, K.: Über den Argininstoffwechsel. II. Mitteilung. Z. physiol. Chem. **178**, 192—201 (1928).
FELIX, K., INOUYE, K., DIRR, K.: (1) Über Clupein. V. Mitteilung. Z. physiol. Chem. **211**, 187—202 (1932).
FELIX, K., DIRR, K., HOFF, A.: (2) Über Clupein. VI. Mitteilung. Z. physiol. Chem. **212**, 50—52 (1932).
FELIX, K., HIROHATA, R., DIRR, K.: Über Clupein. VII. Mitteilung. Z. physiol. Chem. **218**, 269—279 (1933).
FELIX, K., BAUMER, L., SCHÖRNER, E.: Über das Schicksal der Protamine in der befruchteten Eizelle. Z. physiol. Chem. **243**, 43—56 (1936).
FELIX, K., FISCHER, H., KREKELS, A., RAUEN, H. M.: Über Clupein. IX. Mitteilung. Z. physiol. Chem. **286**, 67—78 (1950).
FELIX, K., FISCHER, H., KREKELS, A., MOHR, R.: (1) Nucleoprotamin. I. Mitteilung. Z. physiol. Chem. **287**, 224—234 (1951).
FELIX, K., FISCHER, H., KREKELS, A., MOHR, R.: (2) Nucleoprotamin. II. Mitteilung. Z. physiol. Chem. **289**, 10—19 (1951).
FELIX, K., FISCHER, H., KREKELS, A.: (1) Nucleoprotamin. III. Mitteilung. Z. physiol. Chem. **289**, 127—131 (1952).
FELIX, K., HARTLEIB, J., KREKELS, A.: (2) Nucleoprotamin. IV. Mitteilung. Z. physiol. Chem. **290**, 66—69 (1952).
FELIX, K., RAUEN, H. M., ZIMMER, G. H.: (3) Über Clupein. XI. Mitteilung. Z. physiol. Chem. **291**, 228—234 (1952).
FELIX, K., FISCHER, H., KREKELS, A.: Protamines and nucleoprotamines. Progr. Biophys. Biophys. Chem. **6**, 1—23 (1956).
FELIX, K., KREKELS, A., RICK, W.: The structure of protamines and their formation during spermatogenesis. Trans. Faraday Soc. **53**, 252—253 (1957).
FELIX, K., GOPPOLD-KREKELS, A., LEHMANN, H.: Die Bildung der Nucleoprotamine während der Spermiogenese. Z. physiol. Chem. **312**, 57—67 (1958).
FEUGHELMAN, M., LANGRIDGE, R., SEEDS, W. E., STOKES, A. R., WILSON, H. R., HOOPER, C. W., WILKINS, M. H. F., BARCLAY, R. K., HAMILTON, L. D.: Molecular structure of deoxyribose nucleic acid and nucleoprotein. Nature (Lond.) **175**, 834—838 (1955).
FISCHER, H., KREUZER, L.: Über Gallin. Z. physiol. Chem. **293**, 176—182 (1953).
FITCH, W. M.: Evolution of clupeine Z, a probable crossover product. Nature (Lond.) New Biol. **229**, 245—247 (1971).
FRAENKEL-CONRAT, H. L., OLCOTT, H. S.: Possible cyclic structure of salmine. Fed. Proc. **6**, 253 (1947).
GARVIE, W. H. H.: The action of protamine derivatives and nitrogen mustard on the growth of the Walker 256 rat carcinoma. Brit. J. Cancer **19**, 519—526 (1965).
GEHATIA, M., HASHIMOTO, C.: Physikalisch-Chemische Untersuchung der Protamine. I. Teil. Iridin (aus *Salmo irideus*). Biochim. biophys. Acta (Amst.) **69**, 212—221 (1963).
GEIDUSCHEK, E. P.: On the factors controlling the reversibility of DNA denaturation. J. molec. Biol. **4**, 467—487 (1962).
GEORGIEV, G. P., ERMOLAEVA, L. P., ZBARSKII, I. B.: The quantitative relations between protein and nucleoprotein fractions of various tissues. Biokhimiya **25**, 318—322 (1960).
GLEDHILL, B. L., GLEDHILL, M. P., RIGLER, R., Jr., RINGERTZ, N. R.: Changes in deoxyribonucleoprotein during spermiogenesis in the bull. Exp. Cell Res. **41**, 652—665 (1966).
GOPPOLD-KREKELS, A., LEHMANN, H.: Papierchromatographische Aufteilung von Protaminen. Z. physiol. Chem. **313**, 147—151 (1958).

GOTO, M.: Über die Protamine. Z. physiol. Chem. **37**, 94—114 (1902).
GOTO, M.: On protamines (in Japanese). Igaku-Chuo-Zasshi (Japana Centra Revuo Medicina) **3**, 17—19 (1905).
GROSS, R. E.: Ein Beitrag zur Kenntnis der Protamine. Z. physiol. Chem. **120**, 167—184 (1922).
HAGEDORN, H. C., JENSEN, B. N., KRARUP, N. B., WODSTRUP, I.: Protamine insulinate. J. Amer. med. Ass. **106**, 177—180 (1936).
HAGGIS, G. H.: Proton-deuteron exchange in protein and nucleoprotein molecules surrounded by heavy water. Biochim. biophys. Acta (Amst.) **23**, 494—503 (1957).
HAMAGUCHI, K., GEIDUSCHEK, E. P.: The effect of electrolytes on the stability of the deoxyribonucleate helix. J. Amer. chem. Soc. **84**, 1329—1338 (1962).
HAMMARSTEN, E.: Zur Kenntnis der bilogischen Bedeutung der Nucleinsäureverbindungen. Biochem. Z. **144**, 383—466 (1924).
HASHIMOTO, C.: Studies on DNP-protamines by means of the absorption spectra. Bull. Chem. Soc. Japan **28**, 385—389 (1955).
HASHIMOTO, C.: Chemical studies of the nucleoproteins in sperm and testis cell nuclei of fishes. I. A study on the chemical structure of iridine (in Japanese). Nippon Kagaku Zasshi (J. Chem. Soc. Japan, Pure Chem. Sect.) **79**, 848—855 (1958).
HASHIMOTO, C.: (1) Chemical studies of the nucleoproteins in sperm and testis cell nuclei of fishes. IV. Behaviors toward peptic digestion of basic proteins formed during the formation of spermatozoa of the rainbow-trout *(Salmo irideus)* (in Japanese). Nippon Kagaku Zasshi (J. Chem. Soc. Japan, Pure Chem. Sect.) **80**, 441—444 (1959).
HASHIMOTO, C.: (2) Chemical studies of the nucleoproteins in sperm and testis cell nuclei of fishes. V. Notes on the chemical structures of protamines by means of titration curve, ultraviolet absorption spectrum and SAKAGUCHI's reaction (in Japanese). Nippon Kagaku Zasshi (J. Chem. Soc. Japan, Pure Chem. Sect.) **80**, 800—804 (1959).
HEIM, F.: Pharmakologische Wirkungen des Protamins Clupein. Naunyn-Schmiedebergs Arch. exp. path. Pharmak. **202**, 228—235 (1943).
HELLERMAN, L., PERKINS, M. E.: Activation of enzymes. III. The role of metal ions in the activation of arginase. The hydrolysis of arginine induced by certain metal ions with urease. J. biol. Chem. **112**, 175—194 (1935/36).
HENRICKS, D. M., MAYER, D. T.: Isolation and characterization of a basic keratin-like protein from mammalian spermatozoa. Exp. Cell Biol. **40**, 402—412 (1965).
HERRIOTT, R. M., ANSON, M. L., NORTHROP, J. H.: Reaction of enzymes and proteins with mustard gas [bis (β-chloroethyl) sulfide]. J. gen. Physiol. **30**, 185—210 (1946).
HIROHATA, R.: On mugiline β (in Japanese). Seikagaku (J. Japan. Biochem. Soc.) **30**, 661—666 (1958).
HUANG, R.-C. C., BONNER, J.: Histone, a suppressor of chromosomal RNA synthesis. Proc. nat. Acad. Sci. (Wash.) **48**, 1216—1222 (1962).
HUANG, R.-C. C., BONNER, J., MURRAY, K.: Physical and biological properties of soluble nucleohistones. J. molec. Biol. **8**, 54—64 (1964).
HUISKAMP, W.: Über die Eiweißkörper der Thymusdrüse. Z. physiol. Chem. **32**, 145—196 (1901).
IKOMA, T.: (1) Studies on protamines (V). On peptides in a partial hydrolyzate of mugiline β (in Japanese). Seikagaku (J. Japan. Biochem. Soc. **26**, 13—16 (1954).
IKOMA, T.: (2) Studies on protamines (VI). Counter-current distribution of mugiline β (in Japanese). Seikagaku (J. Japan. Biochem. Soc.) **26**, 460—462 (1954).
IMAHORI, K., TANAKA, J.: Ultraviolet absorption spectra of poly-L-glutamic acid. J. molec. Biol. **1**, 359—364 (1959).
INGLES, C. J., DIXON, G. H.: Phosphorylation of protamine during spermatogenesis in trout testis. Proc. nat. Acad. Sci. (Wash.) **58**, 1011—1018 (1967).
INGLES, C. J., TREVITHICK, J. R., SMITH, M., DIXON, G. H.: Biosynthesis of protamine during spermatogenesis in salmonoid fish. Biochem. biophys. Res. Commun. **22**, 627—634 (1966).
INOUE, S., ANDO, T.: Interaction of clupeine and DNA. Biochim. biophys. Acta (Amst.) **129**, 649—651 (1966).
INOUE, S., ANDO, T.: Optical rotatory dispersion of nucleoclupeine and other model nucleoproteins. Biochem. biophys. Res. Commun. **32**, 501—506 (1968).

INOUE, S., ANDO, T.: Structural characteristics of protamines as observed by their formation of complexes with DNA (in Japanese). Presented at 20th Symposium on Protein Structure, Osaka: Proc. pp. 31—33 (Oct. 1969).

INOUE, S., ANDO, T.: (1) Interaction of clupeine with deoxyribonucleic acid. I. Thermal melting and sedimentation studies. Biochemistry **9**, 388—394 (1970).

INOUE, S., ANDO, T.: (2) Interaction of clupeine with deoxyribonucleic acid. II. Optical rotatory dispersion studies. Biochemistry **9**, 395—396 (1970).

INOUE, S., FUKE, M.: An electron microscope study of deoxyribonucleoprotamines. Biochim. biophys. Acta **204**, 296—303 (1970).

ISHII, S.: Ion-exchange chromatography of amino acids and peptides. I. A new determination method of basic amino acids on a carboxylic acid resin. J. Biochem. (Tokyo) **43**, 531—537 (1956).

ISHII, S.: (1) SAKAGUCHI reaction on arginine-containing peptides (in Japanese). Nippon Kagaku Zasshi (J. Chem. Soc. Japan, Pure Chem. Sect.) **81**, 1586—1588 (1960).

ISHII, S.: (2) The behavior of clupeine toward a carboxylic ion-exchange resin (in Japanese). Nippon Kagaku Zasshi (J. Chem. Soc. Japan, Pure Chem. Sect.) **81**, 1341—1344 (1960).

ISHII, S., YAMASAKI, M., ANDO, T.: Studies on protamines. XI. Peptides obtained by tryptic digestion of clupeine. J. Biochem. (Tokyo) **61**, 687—702 (1967).

ISO, K., KITAMURA, T., WATANABE, I.: Molecular weight of protamines (in Japanese). Nippon Kagaku Zasshi (J. Chem. Soc. Japan, Pure Chem. Sect.) **75**, 342—345 (1954).

ITANO, H. A., GOTTLIEB, A. J.: Blocking of tryptic cleavage of arginyl bonds by the chemical modification of the guanido group with benzil. Biochim. biophys. Res. Commun. **12**, 405—408 (1963).

ITO, H., ICHIKIZAKI, I., ANDO, T.: Synthesis and properties of arginine oligopeptides. Int. J. Protein Res. **2**, 59—65 (1970).

IWAI, K.: The specific degradation of protamine following N-peptidyl to O-peptidyl shift. I. N,O-Peptidyl shift of clupeine by the action of concentrated sulfuric acid (in Japanese). Nippon Kagaku Zasshi (J. Chem. Soc. Japan, Pure Chem. Sect.) **80**, 1066—1070 (1959).

IWAI, K.: The specific degradation of protamine following N-peptidyl to O-peptidyl shift. II. Selective fission of O-peptidyl linkages in the rearranged clupeine (in Japanese). Nippon Kagaku Zasshi (J. Chem. Soc. Japan, Pure Chem. Sect.) **81**, 1302—1307 (1960).

IWAI, K.: The specific degradation of protamine following N-peptidyl to O-peptidyl shift. III. The determination of sequence containing hydroxyamino acid in clupeine and salmine molecules by making use of N → O peptidyl shift (in Japanese). Nippon Kagaku Zasshi (J. Chem. Soc. Japan, Pure Chem. Sect.) **82**, 1088—1096 (1961).

IWAI, K., ANDO, T.: N → O Acyl rearrangement. In: Methods in enzymology (COLOWICK, S. P., KAPLAN, N. O., Eds.). Vol. 11, 263—282. New York-London: Acad. Press. Inc. 1967.

IWAI, K., NAKAHARA, C., ANDO, T.: Studies on protamine. XV. The complete amino acid sequence of the Z component of clupeine. Application of N → O acyl rearrangement and selective hydrolysis in sequence determination. J. Biochem. (Tokyo) **69**, 493—509 (1971).

JAQUES, L. B.: A study of the toxicity of the protamine, salmine. Brit. J. Pharmacol. **4**, 135—144 (1949).

JERGIL, B., DIXON, G. H.: Protamine kinase from rainbow trout testis. Partial purification and characterization. J. biol. Chem. **245**, 425—434 (1970).

JOHNS, E. W., BUTLER, J. A. V.: Specificity of the interactions between histones and deoxyribonucleic acid. Nature (Lond.) **204**, 853—855 (1964).

JUKES, T. H., SCHMIDT, C. L. A.: The combination of certain fatty acids with lysine, arginine, and salmine. J. biol. Chem. **110**, 9—16 (1935).

KAWADE, Y., WATANABE, I.: Sedimentation study of sodium deoxypentosenucleate preparation from herring sperm and calf thymus. Biochim. biophys. Acta (Amst.) **19**, 513—523 (1956).

KAWANISHI, Y., IWAI, K., ANDO, T.: Determination of C-terminal arginine and asparagine of proteins by catalytic hydrazinolysis. J. Biochem. (Tokyo) **56**, 314—324 (1964).

KAWASHIMA, S.: The deoxyribonucleoproteins of fish sperm nuclei. Dissert. for a degree of Dr. Sci., Univ. Tokyo, Dec. 1970.

KAWASHIMA, S., ANDO, T.: Non-basic proteins in sperm cell nuclei of Pacific herring (in Japanese). Presented at 42nd General Meeting of Japan. Biochem. Soc. (Hiroshima, Oct. 8, 1969). Abstr. in Seikagaku (J. Japan. Biochem. Soc.) **41**, 450 (1969). To be published.

KAWASHIMA, S., INOUE, S., ANDO, T.: Interaction of basic oligo-L-amino acids with deoxyribonucleic acid. Oligo-L-ornithines of various chain lengths and herring sperm DNA. Biochim. biophys. Acta (Amst.) **186**, 145—157 (1969).

KAYE, J. S., MCMASTER-KAYE, R.: The fine structure and chemical composition of nuclei during spermiogenesis in the house cricket. I. Initial stages of differentiation and the loss of nonhistone protein. J. Cell Biol. **31**, 159—179 (1966).

KIMURA, M.: (1) The reaction of clupeine and salmine with nitrous acid: The reactivity of N-terminal and guanidino groups (in Japanese). Nippon Kagaku Zasshi (J. Chem. Soc. Japan, Pure Chem. Sect.) **81**, 1154—1157 (1960).

KIMURA, M.: (2) The reaction of protamines with hypobromites (in Japanese). Nippon Kagaku Zasshi (J. Chem. Soc. Japan, Pure Chem. Sect.) **81**, 1861—1865 (1960).

KLOTZ, I. M., GRISWOLD, P., GRUEN, D. M.: Infrared spectra of some proteins and related substances. J. Amer. chem. Soc. **71**, 1615—1620 (1945).

KOGA, Y., SUZUKI, H., TOBITA, T., ANDO, T.: Changes in nuclear proteins during spermatogenesis in rainbow-trout (in Japanese). Presented at 42nd General Meeting of Japan. Biochem. Soc. (Hiroshima, Oct. 7, 1969). Abstr. in Seikagaku (J. Japan. Biochem. Soc.) **41**, 441 (1969).

KOSSEL, A.: Über einen peptonartigen Bestandteil des Zellkerns. Z. physiol. Chem. **8**, 511—515 (1884).

KOSSEL, A.: (1) Über die basischen Stoffe des Zellkerns. S.-B. Kgl. Preuss. Akad. Wiss. **18**, 403—408 (1896).

KOSSEL, A.: (2) Über die basischen Stoffe des Zellkerns. Z. physiol. Chem. **22**, 176—187 (1896).

KOSSEL, A.: Über die Constitution der einfachsten Eiweissstoffe. Z. physiol. Chem. **25**, 165—189 (1898).

KOSSEL, A.: The protamines and histones (Translated by W. V. THORPE), London-New York-Toronto: Longmans, Green and Co. 1928. Protamine und Histone, Leipzig u. Wien: Verlag F. Deuticke 1929.

KOSSEL, A., CAMERON, A. T.: Über die freien Amidogruppen der einfachsten Proteine. Z. physiol. Chem. **76**, 457—463 (1912).

KOSSEL, A., GAWRILOW, N.: Weitere Untersuchungen über die freien Amidogruppen der Proteinstoffe. Z. physiol. Chem. **81**, 274—279 (1912).

KOSSEL, A., KENNAWAY, E. L.: Über Nitroclupein. Z. physiol. Chem. **72**, 486—489 (1911).

KOSSEL, A., KUTSCHER, FR.: Beiträge zur Kenntnis der Eiweißkörper. Z. physiol. Chem. **31**, 165—214 (1900).

KOSSEL, A., MATHEWS, A.: Zur Kenntniss der Trypsinwirkung. Z. physiol. Chem. **25**, 190—194 (1898).

KOSSEL, A., PRINGLE, H.: Über Protamine und Histone. Z. physiol. Chem. **49**, 301—321 (1906).

KOSSEL, A., SCHENCK, E. G.: Untersuchungen über die basischen Eiweissstoffe, ein Beitrag zu ihrer Entwicklungsgeschichte. Z. physiol. Chem. **173**, 278—308 (1928).

KOSSEL, A., STAUDT, W.: Zur Kenntnis der basischen Proteine. Z. physiol. Chem. **159**, 172—178 (1926).

KOSSEL, A., STAUDT, W.: Die Gewinnung eines Argininpeptids aus Clupein. Z. physiol. Chem. **170**, 91—105 (1927).

KOSSEL, A., WEISS, F.: (1) Über die Einwirkung von Alkalien auf Proteinstoffe. I. Mitteilung. Z. physiol. Chem. **59**, 492—498 (1909).

KOSSEL, A., WEISS, F.: (2) Über die Einwirkung von Alkalien auf Proteinstoffe. II. Mitteilung. Z. physiol. Chem. **60**, 311—316 (1909).

KOSSEL, A., WEISS, F.: Über die Einwirkung von Alkalien auf Proteinstoffe. III. Mitteilung. Z. physiol. Chem. **68**, 165—169 (1910).

KOSSEL, A., WEISS, F.: Über einige Nitroderivate von Proteinen. Z. physiol. Chem. **84**, 1—10 (1913).

KUO, J. F., GREENGARD, P. :(1) Cyclic nucleotide-dependent protein kinases. IV. Widespread occurrence of adenosine 3',5'-monophosphate-dependent protein kinase in various tissues and phyla of the animal kingdom. Proc. nat. Acad. Sci. (Wash.) **64**, 1349—1355 (1969).
KUO, J. F., GREENGARD, P.: (2) Adenosine 3',5'-monophosphate-dependent protein kinase from *Escherichia coli*. J. biol. Chem. **244**, 3417—3419 (1969).
LALLIER, R.: Effects de substances polycationiques sur la détermination embryonnaire de l'oeuf de l'oursin *Paracentrotus lividus*. Compt. Rend. **256**, 5409—5412 (1963).
LANGAN, T. A.: Histone phosphorylation: Stimulation by adenosine 3',5'-monophosphate. Science **162**, 579—580 (1968).
LANGAN, T. A., SMITH, L. K.: Phosphorylation of histones and protamines by a specific protein kinase from liver. Fed. Proc. **26**, 603 (1967).
LARSEN, B.: Increased permeability to albumin induced with protamine in modified gelatin membranes. Nature (Lond.) **215**, 641—642 (1967).
LEITCH, G. J., TOBIAS, J. M.: Phospholipid-cholesterol membrane-model; effects of calcium, potassium, or protamine on membrane hydration, water permeability and electrical resistance. J. cell. comp. Physiol. **63**, 225—232 (1964).
LENG, M., FELSENFELD, G.: The preferential interaction of polylysine and polyarginine with specific base sequences in DNA. Proc. nat. Acad. Sci. (Wash.) **56**, 1325—1332 (1966).
LETT, J. T., PARKINS, G. M., ALEXANDER, P.: Physicochemical changes produced in DNA after alkylation. Arch. Biochem. Biophys. **97**, 80—93 (1962).
LEVIN, Y., BERGER, A., KATCHALSKI, E.: Hydrolysis and transpeptidation of lysine peptides by trypsin. Biochem. J. **63**, 308—316 (1956).
LINDERSTRØM-LANG, K.: Some electrochemical properties of a simple protein. Trans. Faraday Soc. **31**, 324—335 (1935).
LINDERSTRØM-LANG, K.: Exchange reactions between D_2O and proteins or protein models. Acta chem. scand. **10**, 149 (1956).
LING, V., TREVITHICK, J. R., DIXON, G. H.: The biosynthesis of protamine in trout testis. I. Intracellular site of synthesis. Canad. J. Biochem. **47**, 51—60 (1969).
LING, V., DIXON, G. H.: The biosynthesis of protamine in trout testis. II. Polysome patterns and protein synthetic activities during testis maturation. J. biol. Chem. **245**, 3035—3042 (1970).
LINSON, L.: Variation de la basophilie pendant la maturation du spermatozoide chez le rat et sa signification histochimique. Acta histochem. (Jena) **2**, 47—67 (1955).
LISNELL, A., MELLGREN, J.: Effect of heparin, protamine, dicumarol, streptokinase, and ε-aminocaproic acid on the growth of human cells *in vitro*. Acta path. microbiol. scand. **57**, 145—153 (1963).
LUZZATI, V.: The structure of nucleohistones and nucleoprotamines. J. molec. Biol. **7**, 758—759 (1963).
LUZZATI, V., NICOLAIEFF, A.: The structure of nucleohistones and nucleoprotamines. J. molec. Biol. **7**, 142—163 (1963).
MARK, J., BISKIND, G. R.: The increased duration of insulin action by the use of protamine-zinc-insulin in pellet form. Endocrinology **26**, 444—448 (1940).
MARUSHIGE, K., BONNER, J.: Template properties of liver chromatin. J. molec. Biol. **15**, 160—174 (1966).
MARUSHIGE, K., DIXON, G. H.: Developmental changes in chromosomal composition and template activity during spermatogenesis in trout testis. Develop. Biol. **19**, 397—414 (1969).
MARUSHIGE, K., LING, V., DIXON, G. H.: Phosphorylation of chromosomal basic proteins in maturing trout testis. J. biol. Chem. **244**, 5953—5958 (1969).
MATSUBARA, H., SINGER, A., SASAKI, R., JUKES, T. H.: Observation on the specificity of a thermostable bacterial protease "thermolysin". Biochem. biophys. Res. Commun. **21**, 242—247 (1965).
McLAREN, A. D.: Concerning the supposed absorption of ultraviolet energy by the peptide linkage. Acta chem. scand. **3**, 648 (1949).
MEISLER, M. H., LANGAN, T. A.: Characterization of a phosphatase specific for phosphorylated histone and protamine. J. biol. Chem. **244**, 4961—4968 (1969).

MENDECKI, J., WILCZOK, T.: The interaction of basic proteins during the donor DNA incorporation into neoplastic cells. Neoplasma (Bratisl.) **10**, 561—564 (1963).

MESSINEO, L.: Modification of the SAKAGUCHI reaction: Spectrophotometric determination of arginine in proteins without previous hydrolysis. Arch. Biochem. Biophys. **117**, 534—540 (1966).

MIESCHER, F.: Über die chemische Zusammensetzung der Eiterzellen. Hoppe-Seilers Med. Chem. Untersuchungen, Berlin **4**, 441, 452 (1870).

MIESCHER, F.: Das Protamin, eine neue organische Base aus den Samenfäden des Rheinlachses. Ber. **7**, 376—379 (1874).

MIESCHER, F.: Die histochemischen und physiologischen Arbeiten, Bd. I und II. Leipzig: F. C. W. Vogel 1897.

MIRSKY, A. E.: Chemical properties of isolated chromosomes. Cold Spr. Harb. Symp. quant. Biol. **12**, 143—146 (1947).

MIRSKY, A. E., POLLISTER, A. W.: Chromosin, a desoxyribose nucleoprotein complex of the cell nucleus. J. gen. Physiol. **30**, 117—148 (1946).

MIRSKY, A. E., RIS, H.: (1) Isolated chromosomes. J. gen. Physiol. **31**, 1—6 (1947).

MIRSKY, A. E., RIS, H.: (2) The chemical composition of isolated chromosomes. J. gen. Physiol. **31**, 7—18 (1947).

MIZUSHIMA, S., SHIMANOUCHI, T., TSUBOI, M.: Near infra-red spectra of proteins and related substances. Nature (Lond.) **166**, 406—407 (1950).

MIYAKE, S.: Die isoelektrischen Punkte der Protamine. Z. physiol. Chem. **172**, 225—229 (1927).

MIYAMOTO, E., KUO, J. F., GREENGARD, P.: (1) Adenosine 3',5'-monophosphate-dependent protein kinase from brain. Science **165**, 63—65 (1969).

MIYAMOTO, E., KUO, J. F., GREENGARD, P.: (2) Cyclic nucleotide-dependent protein kinases. III. Purification and properties of adenosine 3',5'-monophosphate-dependent protein kinase from bovine brain. J. biol. Chem. **244**, 6395—6402 (1969).

MONESI, V.: Autoradiographic evidence of a nuclear histone synthesis during mouse spermiogenesis in the absence of detectable quantities of nuclear ribonucleic acid. Exp. Cell Res. **36**, 683—688 (1964).

MONESI, V.: Synthetic activities during spermatogenesis in the mouse RNA and protein. Exp. Cell Res. **39**, 197—224 (1965)

MONIER, R., JUTISZ, M.: (1) Contribution à l'étude de la structure de la salmine d'*Oncorhynchus*. I. Enchainement des aminoacides au voisinage du résidu N-terminal et étude de quelques peptides résultant de l'hydrolyse acide partiellie. Biochim. biophys. Acta (Amst.) **14**, 551—558 (1954).

MONIER, R., JUTISZ, M.: (2) Contribution à l'étude de la structure de la salmine d'*Oncorhynchus*. II. Étude de quelques peptides résultant de l'hydrolyse trypsique. Biochim. biophys. Acta (Amst.) **15**, 62—68 (1954).

MOORE, S., STEIN, W. H.: Procedures for the chromatographic determination of amino acids on four per cent cross-linked sulfonated polystyrene resins. J. biol. Chem. **211**, 893—906 (1954).

MORISAWA, S.: Studies on protamines. 11. On mugiline β (in Japanese). Fukuoka Igaku Zasshi (Fukuoka Acta Medica) **48**, 427—433 (1957).

MURRAY, K.: Histone nomenclature. In: The nucleohistones (BONNER, J., Ts'o, P. O. P., Eds.), pp. 15—20. San Francisco: Holden-Day Inc. 1964.

MURRAY, K., MILSTEIN, C.: Esters of serine and threonine in hydrolysates of histones and protamines, and attendant errors in amino acid analysis of proteins. Biochem. J. **105**, 491—495 (1967).

MURRAY, K., RASMUSSEN, P. S., NEUSTAEDTER, J., LUCK, J. M.: The hydrolysis of arginine. J. biol. Chem. **240**, 705—709 (1965).

NAKAHARA, C., IWAI, K., ANDO, T.: Fractionation of arginine-containing peptides obtained from clupeine Z component (in Japanese). Seikagaku (J. Japan. Biochem. Soc.) **39**, 463—472 (1967).

NAKANO, M., TOBITA, T., ANDO, T.: A basic protein, galline, from fowl sperm nuclei (in Japanese). Presented at 41st General Meeting of Japan. Biochem. Soc. (Tokyo, Oct. 28, 1968). Abstr. in Seikagaku (J. Japan. Biochem. Soc.) **40**, 555 (1968).

NAKANO, M., TOBITA, T., ANDO, T.: A protamine, galline, from fowl sperm nuclei (in Japanese). Presented at 42nd General Meeting of Japan. Biochem. Soc. (Hiroshima, Oct. 7, 1969). Abstr. in Seikagaku (J. Japan. Biochem. Soc.) **41**, 441 (1969).

NAKANO, M., TOBITA, T., ANDO, T.: Fractionation of galline, a protamine from fowl sperm, and some characterization of the components. Biochim. biophys. Acta (Amst.) **207**, 553—555 (1970).

NAKANO, M., TOBITA, T., ANDO, T.: Characterization of a protamine from fowl sperm (galline) and the primary structure of some components (in Japanese). Proc. 23th Symposium on Protein Structure, Maebashi, Nov. 21—22, pp. 29—32, 1972.

NAKANO, M., TOBITA, T., ANDO, T.: Studies on a protamine (galline) from fowl sperm. 1. Fractionation and some characterization. Int. J. Peptide Protein Res. **5** (1973) (in press).

NELSON-GERHARDT, M.: Untersuchungen über Salmin. Z. physiol. Chem. **105**, 265—282 (1919).

NUKUSHINA, M.: Studies on the chemical structure of clupeine from North-European herrings (in Japanese). Dissert. for a degree of B. Sci., Univ. Tokyo (March 1964).

NUKUSHINA, M., ISHII, S., ANDO, T.: Studies on clupeine obtained from North-European herrings, *Clupea harengus* (in Japanese). Presented at the meeting of Kanto Branch of Japan. Biochem. Soc. (Maebashi, May 9, 1964).

NYGAARD, A. P.: D- and L-Lactic cytochrome c reductase of yeast — interaction of multivalent electrolytes at the acceptor sites. J. biol. Chem. **237**, 742—745 (1962).

OHBA, Y.: Structure of nucleohistone. II. Thermal denaturation. Biochim. biophys. Acta (Amst.) **123**, 84—90 (1966).

OKUYAMA, T., SATAKE, K.: On the preparation and properties of 2,4,6-trinitrophenyl-amino acids and -peptides. J. Biochem. (Tokyo) **47**, 454—466 (1960).

OLINS, D. E., OLINS, A. L., VON HIPPEL, P. H.: Model nucleoprotein complexes: Studies on the interaction of cationic homopolypeptides with DNA. J. molec. Biol. **24**, 157—176 (1967).

OLINS, D. E., OLINS, A. L., VON HIPPEL, P. H.: On the structure and stability of DNA protamine and DNA polypeptide complexes. J. molec. Biol. **33**, 265—281 (1968).

OTA, S.: Studies on protamines (XVII). On the fractionation of mugiline β (in Japanese). Seikagaku (J. Japan. Biochem. Soc.) **33**, 320—324 (1961).

OTA, S., ONOUE, K., HIROHATA, R., KAWACHI, T., OKUDA, Y., MORISAWA, S., FUJII, S.: Über Mugilin β. Z. physiol. Chem. **317**, 1—9 (1959).

OTA, S., MURAMATSU, M., HIROHATA, R., OKUDA, Y., YANG, C.-C., KAO, K.-C., CHIN, W.-C.f CHANG, C.-C., IMAI, Y., ONO, T.: Studies on protamines. XVIII. On the heterogeneity ol mugiline β (in Japanese). Ann. Repts. Lab. Protein Chem., Yamaguchi Med. Schoo No. 1, 1—10 (1966).

PALAU, J., SUBIRANA, J. A.: Histones of marine invertebrates. Biochem. J. **101**, 34 P—35 P (1967).

PARDON, J. F., WILKINS, M. H. F., RICHARDS, B. M.: Super-helical model for nucleohistone. Nature (Lond.) **215**, 508—509 (1967).

PAYZA, A. N., EIBER, H. B., WALTERS, S.: State of tissue lipases after injection of heparin. Proc. Soc. exp. Biol. (N. Y.) **125**, 188—192 (1967).

PHILIPPS, G. R., WEST, J.: Purification of protamine sulfate for the quantitative precipitation of ribonucleic acids. Biochim. biophys. Acta (Amst.) **91**, 416—420 (1964).

PICCARD, J.: Über Protamin, Guanin und Sarkin, als Bestandteile des Lachssperma. Ber. **7**, 1714—1719 (1874).

POLLISTER, A. W., MIRSKY, A. E.: The nucleoprotamine of trout sperm. J. gen. Physiol. **30**, 101—116 (1946).

PORTER, R. R., SANGER, F.: The free amino groups of haemoglobins. Biochem. J. **42**, 287—294 (1948).

PORTIS, E. A., ALTMAN, K. L.: The action of proteolytic enzymes and protaminase on salmine sulfate. J. biol. Chem. **169**, 203—209 (1947).

PRZYLECKI, ST. J. v., MAJMIN, R.: (1) Polysaccharoclupeine. Biochem. Z. **277**, 420—423 (1935).

PRZYLECKI, ST. J. v., GIEDROYĆ, W., RAFALOWSKA, H.: (2) Über den Glykogenzustand im Zellinnern. I. Mitteilung. Über Dreikomponenten-Symplexe aus Clupein, Nucleinsäure, Glykogen oder Dextrin. Biochem. Z. **280**, 286—292 (1935).

PRZYLECKI, ST. J. v., KASPRZYK, K., RAFALOWSKA, H.: Untersuchungen über Polysaccharoproteide. X. Mitteilung: Rolle der einzelnen Aminosäurereste und Schlussbetrachtungen. Biochem. Z. **286**, 360—372 (1936).

RASMUSSEN, K. E.: Clupeinuntersuchungen. I. Darstellung und Fraktionierung von Clupein. Z. physiol. Chem. **224**, 97—115 (1934).

RASMUSSEN, K. E., LINDERSTRØM-LANG, K.: Clupeinuntersuchungen. II. Elektrometrische Titration von Clupein. Z. physiol. Chem. **227**, 181—212 (1934).

RASMUSSEN, P. S.: Chromatographic fractionation of protamine on Amberlite IRC-50 columns with guanidinium chloride. Presented at XI. Scand. Physiol. Congr. (Copenhagen, 1963). Abstr. in Acta physiol. scand. **59**, Suppl. No. 213, 146 (1963).

RAUEN, H. M., STAMM, W., FELIX, K.: Über Clupein. XII. Mitteilung. Z. physiol. Chem. **291**, 275—278 (1952).

RAUEN, H. M., STAMM, W., FELIX, K.: Gegenstromverteilung von Protaminen. Z. physiol. Chem. **292**, 101—109 (1953).

RAUKAS, E.: Melting point of DNA complexes with protamines, basic polypeptides, and polyethylenepolyamine. Biokhimiya **30**, 1122—1131 (1965).

RAUKAS, E.: Thermal denaturation of DNA complexes with protamine peptides. Eesti NSV Teaduste Akad. Toimetised, Biol. Seer. **15**, 342—346 (1966).

RAUKAS, E., STRUCHKOV, V., STRAZHEVSKAYA, N.: (1) The micellar structure of DNA. Eesti NSV Teaduste Akad. Toimetised, Biol. Seer. **15**, 161—169 (1966).

RAUKAS, E., MITYUSHIN, V., KAFTANOVA, A.: (2) Structure of deoxyribonucleoprotein of sperm nuclei: Nucleoprotamine structure. Eesti NSV Teaduste Akad. Toimetised, Biol. Seer. **15**, 467—479 (1966).

ROCHE, J., BLANC-JEAN, G.: Sur l'état des groupments guanidiques dans les molécules protéiques. Compt. rend. **210**, 681—683 (1940).

ROCHE, J., MOURQUE, M.: Nitration des protéines réactivité des groupments guanidiques de l'arginine. Compt. rend. **226**, 1848—1850 (1948).

ROGOZINSKI, F.: Über die Einwirkung von proteolytischen Fermenten auf Clupein. Z. physiol. Chem. **79**, 398—414 (1912).

ROSENHECK, K., DOTY, P.: The far ultraviolet absorption spectra of polypeptide and protein solutions and their dependence on conformation. Proc. nat. Acad. Sci. (Wash.) **47**, 1775—1791 (1961).

ROSS, V.: (1) Hydrogen-ion activity and precipitation of bovine serum albumin and β-lactoglobulin with salmine. Arch. Biochem. Biophys. **50**, 34—45 (1954).

ROSS, V.: (2) Reaction of salmine with ovalbumin, hemoglobin, insulin and casein. Arch. Biochem. Biophys. **50**, 46—54 (1954).

SAKAGUCHI, S.: A new method for the colorimetric determination of arginine. J. Biochem. (Tokyo) **37**, 231—236 (1950).

SAKAGUCHI, S.: Note to the colorimetric determination of arginine. J. Biochem. (Tokyo) **38**, 91 (1951).

SANGER, F.: The free amino groups of insulin. Biochem. J. **39**, 507—515 (1945).

SANGER, F.: The structure of insulin. In: Currents in biochemical research (GREEN, E., Ed.), pp. 434—459. New York: John Wiley 1956.

SANWAL, B. D., MAEBA, P., COOK, R. A.: Interaction of macroions and dioxane with the allosteric phosphoenolpyruvate carboxylase. J. biol. Chem. **241**, 5177—5182 (1966).

SAWADA, F.: The fundamental conditions for countercurrent distribution studies of clupeine. Bull. Chem. Soc. Japan **32**, 161—165 (1959).

SAWADA, F., ANDO, T.: Studies on the origin of heterogeneity in protamines by means of chromatographic method. Presented at 31st General Meeting of Japan. Biochem. Soc. (Sapporo, July, 1958). Abstr. in Seikagaku (J. Japan. Biochem. Soc.) **30**, 808 (1958).

SCANES, F. S., TOZER, B. T.: Fractionation of basic proteins and polypeptides. Clupeine and Salmine. Biochem. J. **63**, 565—576 (1956).

SCHENCK, E. G.: Über die Beeinflussbarkeit der Blutzuckerregulation durch Eiweissstoffe, Aminosäuren und deren Derivate. I. Naunyn-Schmiedebergs Arch. exp. Path. Pharmak. **167**, 201—215 (1932).

SCHLOSSMAN, S. F., YARON, A., BEN-EFRAIM, S., SOBER, H. A.: Immunogenicity of a series of α-N-DNP-L-lysines. Biochemistry **4**, 1638—1645 (1965).

SCHNEIDER, R.: Nucleoprotamin. VI. Mitteilung. Karyometrische Untersuchungen an durch "künstliche Befruchtung" entstandenen Bachsaibling-Embryonen. Z. physiol. Chem. **294**, 74—79 (1954).

SKALKA, A., FOWLER, A. V., HURWITZ, J.: The effect of histones on the enzymatic synthesis of ribonucleic acid. J. biol. Chem. **241**, 588—596 (1966).

SMULL, C. E., LUDWIG, E. H.: Enhancement of the plaque-forming capacity of polio virus ribonucleic acid with basic proteins. J. Bact. **84**, 1035—1040 (1962).

SMULL, C. E., MALLETTE, M. F., LUDWIG, E. H.: The use of basic proteins to increase infectivity of enterovirus ribonucleic acid. Biochem. biophys. Res. Commun. **5**, 247—249 (1961).

ŠORM, F., ŠORMOVA, Z.: On proteins and amino-acids. VII. On clupein. Collect. Czech. Chem. Commun. **16**, 207—213 (1951).

SPACKMAN, D. H., STEIN, W. H., MOORE, S.: Automatic recording apparatus for use in the chromatography of amino acids. Anal. Chem. **30**, 1190—1206 (1958).

STEDMAN, E., STEDMAN, E.: (1) The function of deoxyribose-nucleic acid in the cell nucleus. Symp. Soc. exp. Biol. **1**, 232—251 (1947a).

STEDMAN, E., STEDMAN, E.: (2) The chemical nature and functions of the components of cell nuclei. Cold Spr. Harb. Symp. quant. Biol. **12**, 224—236 (1947).

STEDMAN, E., STEDMAN, E.: Cell specificity of histones. Nature (Lond.) **166**, 780—781 (1950).

STEDMAN, E., STEDMAN, E.: The basic proteins of cell nuclei. Phil. Trans. B **235**, 565—596 (1951).

STEUDEL, H.: Zur Histochemie der Spermatozoen. III. Mitteilung. Z. physiol. Chem. **83**, 72—78 (1913).

STEUDEL, H., PEISER, E.: Über Nucleinsäure-Eiweissverbindungen. Z. physiol. Chem. **122**, 298—306 (1922).

SUZUKI, KE.: On the formation of the precipitate of nucleoclupein from sodium desoxypentosenucleate and clupein sulphate (in Japanese). Rep. Inst. Sci. and Techn., Univ. Tokyo **4**, 231—236 (1950).

SUZUKI, KO., ANDO, T.: (1) Studies on the conformation of clupeine. J. Biochem. (Tokyo) **63**, 403—405 (1968).

SUZUKI, KO., ANDO, T.: (2) Studies on protamines. XIII. The fractionation of clupeine Y. J. Biochem. (Tokyo) **63**, 701—708 (1968).

SUZUKI, KO., ANDO, T.: Inhibition of RNA synthesis by clupeine and basic homopolypeptides. J. Biochem. (Tokyo) **65**, 831—834 (1969).

SUZUKI, KO., ANDO, T.: (1) Studies on protamines. XVI. The complete amino acid sequence of clupeine YII. J. Biochem. (Tokyo) **72**, 1419—1432 (1972).

SUZUKI, KO., ANDO, T.: (2) Studies on protamines. XVII. The complete amino acid sequence of clupeine YI. J. Biochem. (Tokyo) **72**, 1433—1446 (1972).

SUZUKI, KE., WATANABE, I.: (1) Studies on nucleosalmine. I. Extraction, purification, and properties (in Japanese). Nippon Kagaku Zasshi (J. Chem. Soc. Japan, Pure Chem. Sect.) **73**, 778—781 (1952).

SUZUKI, KE., WATANABE, I.: (2) Studies on nucleosalmine. II. Formation of artificial nucleoproteins and the influence by pH (in Japanese). Nippon Kagaku Zasshi (J. Chem. Soc. Japan, Pure Chem. Sect.) **73**, 825—827 (1952).

SUZUKI, KE., WATANABE, I.: On pentosenucleic acid-protamine complexes (in Japanese). Nippon Kagaku Zasshi (J. Chem. Soc. Japan, Pure Chem. Sect.) **74**, 689—692 (1953).

THOMPSON, W. H.: Die physiologische Wirkung der Protamine und ihrer Spaltungsprodukte. Z. physiol. Chem. **29**, 1—19 (1900).

TOBITA, T., YAMASAKI, M., ANDO, T.: Studies on protamines. XII. Determination of the carboxyl-terminal structure of clupeine and salmine using enzymatic procedures. J. Biochem. (Tokyo) **63**, 119—126 (1968).

TREVITHICK, J. R., INGLES, C. J., DIXON, G. H.: Biosynthesis of protamine in trout testis. Fed. Proc. **26**, 603 (1967).

TRISTRAM, G. R.: Constitution of salmine. Nature (Lond.) **160**, 637 (1947).

TRISTRAM, G. R.: Chapt. VII. Amino acid composition of certain proteins. Advanc. Protein Chem. **5**, 129—141 (1949).

TSURU, D., KIRA, H., YAMAMOTO, T., FUKUMOTO, J.: Studies on bacterial protease. Part XVIII. Proteolytic specificity of neutral protease of *Bacillus subtilis var. amylosacchariticus*. Agr. Biol. Chem. (Japan) **31**, 718—723 (1967).
UI, N.: The influence of association phenomenon of salmine on the electrophoretic pattern (Studies on protamines by means of electrophoresis. III.) (in Japanese). Nippon Kagaku Zasshi (J. Chem. Soc. Japan, Pure Chem. Sect.) **77**, 947—951 (1956).
UI, N., WATANABE, I.: Studies on protamines by means of electrophoresis (I. and II.). I. Electrophoretic mobilities and electric charges of protamines; II. Combination of anions with salmine. Nippon Kagaku Zasshi (J. Chem. Soc. Japan, Pure Chem. Sect.) **74**, 647—651, 651—653 (1953) (in Japanese).
VAUGHN, J. C.: The relationship of the "sphere chromatophile" to the fate of displaced histones following histone transition in rat spermiogenesis. J. Cell Biol. **31**, 257—278 (1966).
VELICK, S. F., UDENFRIEND, S.: The amino end-group and the amino acid composition of salmine. J. biol. Chem. **191**, 233—238 (1951).
WALDSCHMIDT-LEITZ, E.: (1) Über die Struktur der einfachsten Eiweisskörper. Monatsh. **66**, 357—366 (1935).
WALDSCHMIDT-LEITZ, E., GUDERNATSCH, H.: Über die Struktur der Protamine. V. Über die Beziehungen zwischen Zusammensetzung und Reifegrad des Clupeins. Z. physiol. Chem. **309**, 266—275 (1957).
WALDSCHMIDT-LEITZ, E., GUTERMANN, H.: Über die Struktur der Protamine. VI. Vergleich der Protamine aus Salmonidenarten. Z. physiol. Chem. **323**, 98—104 (1961).
WALDSCHMIDT-LEITZ, E., GUTERMANN, H.: Über die Struktur der Protamine. VII. Über den Einfluss des Alters von Fischen auf die Zusammensetzung ihrer Protamine. Z. physiol. Chem. **344**, 50—54 (1966).
WALDSCHMIDT-LEITZ, E., KOFRANYI, E.: Darstellung von Protaminase. Z. physiol. Chem. **222**, 148—150 (1933).
WALDSCHMIDT-LEITZ, E., KOFRANYI, E.: Über die Struktur der Protamine. II. Strukturaufklärung des Clupeins. (Sechste Mitteilung über enzymatische Proteolyse). Z. physiol. Chem. **236**, 181—191 (1935).
WALDSCHMIDT-LEITZ, E., PFLANZ, L.: Über die Struktur der Protamine. III. Dinitrophenylclupein und seine enzymatische Spaltung. Z. physiol. Chem. **292**, 150—156 (1953).
WALDSCHMIDT-LEITZ, E., PURR, A.: (1) Über Proteinase und Carboxy-Polypeptidase aus Pankreas. (XVII. Mitteilung zur Spezifität tierischer Proteasen). Ber. **62**, 2217—2226 (1929).
WALDSCHMIDT-LEITZ, E., VOH, R.: Über die Struktur der Protamine. IV. Fraktionierung von Clupeine. Z. physiol. Chem. **298**, 257—267 (1954).
WALDSCHMIDT-LEITZ, E., SCHÄFFNER, A., GRASSMANN, W.: Über die Struktur des Clupeins. Z. physiol. Chem. **156**, 68—98 (1926).
WALDSCHMIDT-LEITZ, E., STADLER, P., STEIGERWALDT, F.: Über Blutgerinnung. Hemmung und Beschleunigung. Z. physiol. Chem. **183**, 39—59 (1929).
WALDSCHMIDT-LEITZ, E., ZIEGLER, F., SCHÄFFNER, A., WEIL, L.: Über die Struktur der Protamine. I. Protaminase und die Produkte ihrer Einwirkung auf Clupein und Salmin. (Fünfte Mitteilung über enzymatische Proteolyse.) Z. physiol. Chem. **197**, 219—236 (1931).
WALDSCHMIDT-LEITZ, E., KÜHN, K., ZINNERT, F.: Zur Bausteinanalyse des Clupeins. Experientia (Basel) **7**, 183—184 (1951).
WALEY, S. G., WATSON, J.: Rearrangement of the amino-acid residues in peptides by the action of proteolytic enzymes. Nature (Lond.) **167**, 360—361 (1951).
WALSH, D. A., PERKINS, J. P., KREBS, E. G.: An adenosine 3',5'-monophosphate-dependent protein kinase. J. biol. Chem. **243**, 3763—3765 (1968).
WANG, T. Y.: The isolation, properties, and possible functions of chromatin acidic proteins. J. biol. Chem. **242**, 1220—1226 (1967).
WARBURG, O., CHRISTIAN, W.: Isolierung und Kristallisation des Proteins des oxydierenden Gärungsferments. Biochem. Z. **303**, 40—68 (1939).
WARREN, J., WEIL, M. L., RUSS, S. B., JEFFRIES, H.: Purification of certain viruses by use of protamine sulfate. Proc. Soc. exp. Biol. (N. Y.) **72**, 662—664 (1949).

WATANABE, S.: Studies on the chemical structure of salmine and iridine. Dissert. for a degree of Dr. Sci., Univ. Toyko (June, 1969).
WATANABE, I., SUZUKI, KE.: (1, 2) Studies on nucleoclupeine (I. and II.) I. Extraction and purification of nucleoclupeine. II. Separation of desoxypentose-nucleic acid and clupeine (in Japanese). Nippon Kagaku Zasshi (J. Chem. Soc. Japan, Pure Chem. Sect.) **72**, 578—580, 580—583 (1951).
WATANABE, I., SUZUKI, KE.: (3) Studies on nucleoclupeine (III). On the conditions for the formation of fibrous precipitate of nucleoclupeine (in Japanese). Nippon Kagaku Zasshi (J. Chem. Soc. Japan, Pure Chem. Sect.) **72**, 604—606 (1951).
WEBER, C. T.: A modification of SAGAKUCHI reaction for the quantitative determination of arginine. J. biol. Chem. **86**, 217—222 (1930).
WEIL, L., SEIBLES, T. S., TELKA, M.: Specificity of protaminase. Arch. Biochem. Biophys. **79**, 44—54 (1959).
WIGLE, D. T., DIXON, H. H.: Transient incorporation of methionine at the N-terminus of protamine newly synthesized in trout testis cells. Nature (Lond.) **227**, 676—680 (1970).
WILCZOK, T., MENDECKI, J.: The effect of protamines and histones on incorporation of donor DNA into neoplastic cells. Neoplasma (Bratisl.) **10**, 113—119 (1963).
WILKINS, M. H. F.: Physical studies of the molecular structure of deoxyribose nucleic acid and nucleoprotein. Cold Spr. Harb. Symp. quant. Biol. **21**, 75—90 (1956).
WILKINS, M. H. F., RANDALL, J. T.: Crystallinity in sperm heads: Molecular structure of nucleoprotein *in vivo*. Biochim. biophys. Acta (Amst.) **10**, 192—193 (1953).
WILKINS, W. H. F., ZUBAY, G.: X-ray diffraction studies of molecular structure of nucleohistone and chromosomes. J. molec. Biol. **1**, 179—185 (1959).
WILKINS, M. H. F., ZUBAY, G.: X-ray diffraction study of the structure of nucleohistone and nucleoprotamines. J. molec. Biol. **7**, 756—757 (1963).
WILKINS, M. H. F., STOKES, A. R., WILSON, H. R.: Molecular structure of deoxypentose nucleic acids. Nature (Lond.) **171**, 738—740 (1953).
YAMASAKI, M.: Chemical and enzymic studies on protamines. I. Preparation and some characteristics of clupeine and salmine. Sci. Pap. Coll. Gen. Educ., Univ. Tokyo **8**, 165—173 (1958).
YAMASAKI, M.: (1) Chemical and enzymic studies on protamines. II. N-Terminal residues and molecular weights of clupeine and salmine. Sci. Pap. Coll. Gen. Educ., Univ. Tokyo **9**, 31—47 (1959).
YAMASAKI, M.: (2) Chemical and enzymic studies on protamines. III. N-Terminal sequence of clupeine and salmine. Sci. Pap. Coll. Gen. Educ., Univ. Tokyo **9**, 49—65 (1959).
YAMASAKI, M.: Chemical and enzymic studies on protamines. IV. Effects of endopeptidases on clupeine and salmine. Sci. Pap. Coll. Gen. Educ., Univ. Tokyo **10**, 37—47 (1960).
YANG, J. T., DOTY, P.: The optical rotatory dispersion of polypeptides and proteins in relation to configuration. J. Amer. chem. Soc. **79**, 761—775 (1957).
YANKEELOV, J., Jr., KOCHERT, M., PAGE, J., WESTPHAL, A.: A reagent for the modification of arginine residues under mild conditions. Fed. Proc. **25**, 590 (1966).
YOSHIDA, H., FUJISAWA, H., OHI, Y.: Influence of protamine on the Na^+, K^+-dependent ATPase and on the active transport processes of potassium and of L-DOPA into brain slices. Canad. J. Biochem. **43**, 841—849 (1965).
ZICKLER, F.: Die Wirkung freier Phagenrezeptoren auf Proteus-Phagen in Gegenwart von Polykationen. Z. Naturforsch. **22** b, 418—421 (1967).
ZIMMERMANN, E.: Dissert., Johann Wolfgang Goethe-Universität, Frankfurt a. M. (Deutschland) 1959.
ZUBAY, G., DOTY, P.: The isolation and properties of deoxyribonucleoprotein particles containing single nucleic acid molecules. J. molec. Biol. **1**, 1—20 (1959).
ZUBAY, G., WILKINS, M. H. F.: An X-ray diffraction study of histone and protamine in isolation and in combination with DNA. J. molec. Biol. **4**, 444—450 (1962).

Subject Index

Molecular Biology, Biochemistry and Biophysics

Volumes published:

Vol. 1 J. H. van't Hoff: Imagination in Science. Translated into English, with Notes and a General Introduction by G. F. Springer. With 1 portrait. VI, 18 pages. 1967. DM 6,60; US $ 2.80

Vol. 2 K. Freudenberg and A. C. Neish: Constitution and Biosynthesis of Lignin. With 10 figures. IX, 129 pages. 1968. DM 28,—; US $ 11.50

Vol. 3 T. Robinson: The Biochemistry of Alkaloids. With 37 figures. X, 149 pages. 1968. DM 39,—; US $ 16.—

Vol. 4 A. S. Spirin and L. P. Gavrilova: The Ribosome. With 26 figures. X, 161 pages. 1969. DM 54,—; US $ 22.20

Vol. 5 B. Jirgensons: Optical Rotatory Dispersion of Proteins and Other Macromolecules. With 65 figures. XI, 166 pages. 1969. DM 46,—; US $ 18.90

Vol. 6 F. Egami and K. Nakamura: Microbial Ribonucleases. With 5 figures. IX, 90 pages. 1969. DM 28,—; US $ 11.50

Vol. 8 Protein Sequence Determination. A Sourcebook of Methods and Techniques. Edited by Saul B. Needleman. With 77 figures. XXI, 345 pages. 1970. DM 84,—; US $ 34.50

Vol. 9 R. Grubb: The Genetic Markers of Human Immunoglobulins. With 8 figures. XII, 152 pages. 1970. DM 42,—; US $ 17.30

Vol. 10 R. J. Lukens: Chemistry of Fungicidal Action. With 8 figures. XIII, 136 pages. 1971. DM 42,—; US $ 17.30

Vol. 11 P. Reeves: The Bacteriocins. With 9 figures. XI, 142 pages. 1972. DM 48,—; US $ 19.70

Vol. 12 T. Ando, M. Yamasaki and K. Suzuki: Protamines: Isolation, Characterization, Structure and Function. With 24 figures. IX. 114 pages. 1973. DM 48,—; US $ 19.70

Vol. 13 P. Jollès and A. Paraf: Chemical and Biological Basis of Adjuvants. With 24 figures. VIII, 153 pages. 1973. DM 48,—; US $ 19.70

Vol. 14 Micromethods in Molecular Biology. Edited by V. Neuhoff. With 275 figures. XV, 428 pages. 1973. DM 98,—; US $ 40.20

Prices are subject to change without notice

Volumes in preparation:

Breuer, H., and K. Dahm: Mode of Action of Estrogens

Burns, R. C., and R. W. F. Hardy: Nitrogen Fixation in Bacteria and Higher Plants

Callan, H. G., and O. Hess: Lampbrush Chromosomes

Chapman, D.: Molecular Biological Aspects of Lipids in Membranes

Haugaard, N., and E. Haugaard: Mechanism of Action of Insulin

Hawkes, F.: Nucleic Acids and Cytology

Volumes in preparation (continued):

JARDETZKY, O., and G. C. K. ROBERTS: High Resolution Nuclear Magnetic Resonance in Molecular Biology

KERSTEN, W., and H. KERSTEN: Antibiotics Inhibiting Nucleic Acid Synthesis

LAUFFER, M. A.: Entropy-Driven Processes in Biology

MATHEWS, M. B.: Molecular Evolution of Connective Tissue

QUASTEL, J. H., and C. S. SUNG: Chemical Regulatory Processes in the Nervous System

SCHANNE, O., and ELENA RUIZ DE CERETTI: Impedance Measurements in Biological Cells

SHULMAN, S.: Tissue Specificity and Autoimmunity

SÖLL, D., and U. L. RAJBHANDARY: Transfer-RNA

VINNIKOV, J. A.: Cytological and Molecular Basis of Sensory Reception

WEISSBLUTH, M.: Hemoglobin: Cooperativity and Electronic Properties

YAGI, K.: Biochemical Aspects of Flavins

Practical Molecular Genetics. Edited by H. MATTHAEI and F. GROS